Technological Advancement in E-waste Management

The theme of this book is sustainable e-waste management through effective amalgamation of information and communication technologies (ICT) and green recycling technologies to ensure development of intelligent, smart, and sustainable systems. It encompasses multidisciplinary interventions, including recent case studies from OEMs and IT industries as well as e-waste recyclers, and explores interdisciplinary research and industry–academia collaborations towards the development of smart and sustainable systems for e-waste management.

Features:

- Covers the application of smart and intelligent systems for e-waste management.
- Explores recent advancements from the technological aspect – both recycling and ICT.
- Reviews supply chain criticalities for e-waste.
- Aims at cleaner production and intelligent systems for a green digital economy.
- Includes real-life case studies reflecting industry standards and the current paradigm.

This book is aimed at graduate students and researchers in environmental engineering, waste management, urban mining, circular economy, waste processing, electronics and telecommunication engineering, electrical and electronics engineering, and chemical engineering.

Intelligent Data-Driven Technology for Sustainability

Series Editors: Siddhartha Bhattacharyya, Mario Koeppen, Sergey Gorbachev, Khan Muhammad, and Naba Kumar Mondal

The proposed series aims to describe evolve, design, and develop intelligent technology and models for analysis of data involved in environmental engineering with the objective of ensuring sustainability. This series would unveil intelligent and cognitive models to address issues related to effective monitoring of environmental pollution and ushering in a sustainable environmental design, thereby contributing to the overall well-being of the global environment for better sustenance and livelihood. As such, the contributory volumes would remain centered on evolving novel intelligent/cognitive models and algorithms to evolve sustainable solutions for the mitigation of environmental pollution. Sufficient coverage of novel cognitive models for the purpose of effective environmental pollution data management at par with the standards laid down by World Health Organization, is also envisaged.

Technological Advancement in E-waste Management
Towards Smart, Sustainable, and Intelligent Systems
Edited by Biswajit Debnath, Abhijit Das, Potluri Anil Chowdary, and Siddhartha Bhattacharyya

For more information about this series, please visit: www.routledge.com/Intelligent-Data-Driven-Technology-for-Sustainability/book-series/CRCIDTS

Technological Advancement in E-waste Management

Towards Smart, Sustainable, and Intelligent Systems

Edited by
Biswajit Debnath, Abhijit Das,
Potluri Anil Chowdary and
Siddhartha Bhattacharyya

CRC Press
Taylor & Francis Group
Boca Raton London New York

CRC Press is an imprint of the
Taylor & Francis Group, an **informa** business

First edition published 2024
by CRC Press
6000 Broken Sound Parkway NW, Suite 300, Boca Raton, FL 33487-2742

and by CRC Press
4 Park Square, Milton Park, Abingdon, Oxon, OX14 4RN

CRC Press is an imprint of Taylor & Francis Group, LLC

ISBN: 978-1-032-32856-0 (hbk)
ISBN: 978-1-032-32857-7 (pbk)
ISBN: 978-1-003-31705-0 (ebk)

DOI: 10.1201/9781003317050

Typeset in Times
by MPS Limited, Dehradun

Dedication

Biswajit would like to dedicate this book to his two mothers
Mrs. Rekha Debnath and Mrs. Ratna Das (Mauma)

Abhijit would like to dedicate this book to his Mother
Mrs. Monika Das and his father Mr. Gour Hari Das

Anil would like to dedicate this book to his sister Dr. K. Sushma,
his mother P. Manjula and his wife Brahmani

Siddhartha would like to dedicate this volume to his fellow editors
who have the habit to make impossible a possible

Contents

Preface

E-waste is a mini catastrophe and a global threat to the overall anthropogenosphere. The huge amount of e-waste is a result of technological advancement coupled with short innovation cycles and intelligent marketing strategies that shorten the lifespan of the electronic item. Globally, e-waste is the fastest expanding waste stream in the world, increasing at an annual rate of 3–5%. There has been an acceleration in research and development of environmentally sound e-waste recycling technologies in the past decade. E-waste is a rich source of metals, glass fibre, polymers, etc., which makes it the best potential candidate for urban mining. This huge resource present in the molecules and networks of e-waste would be in vain unless tapped and brought back to the economy. Recent trends in recycling technology development are focused on valorization routes rather than just recovering. The approaches taken by the researchers are unique and greener. For example, utilization of green solvents for recovery, green catalyst development from e-waste, green pyrolysis of e-waste, etc. Hence, the focus is on developing sustainable technologies for green urban mining.

While there is one branch of research highly focused on recycling technologies, it requires rigorous wet-laboratory hours; the researchers in the dry-laboratory are also brainstorming to solve this problem using technological innovations in their forte. It is a popular opinion that waste management is a highly interdisciplinary area, and e-waste is no different. In the last few years, information and communication technologies (ICT) are slowly sweeping into this sector, which is allowing the development of new approaches and strategies for e-waste management. One important aspect of e-waste management is supply chain management. The collection to disposal mechanism of e-waste is a complex one due to the presence of so many actors and players, including the informal sector. It has caught the attention of researchers in recent years. A quick analysis reveals that subject areas, such as machine learning, game theory, fuzzy theory, cloud computing, Internet of Things (IoT), etc., are going global to tackle e-waste. Additionally, blockchain technology is revolutionizing the supply chain area for its unique features such as traceability – not to mention that these technologies have the potential to contribute towards the development of smart and intelligent systems.

E-waste management is a complex and inter-disciplinary area. Technological proliferations in the area, such as IoT, blockchain technologies, machine learning, etc., will increase the benchmark for e-waste management systems towards smart and intelligent systems. The progress in recycling and valorization technologies, including nanotechnology, green catalyst development, etc., will dictate a cleaner production regime. Concepts, such as smart city, green economy, sustainable city, etc., can be of additional advantage to amalgamate all these technological advancements towards the development of sustainable systems. In this view, there are three major tracks in this book, divided into several chapters.

Part 1 presents the overview of e-waste management technologies with sustainability on focus. This includes both recycling technologies as well as digital technologies.

Part 2 disseminates works on e-waste that contribute towards the proliferation of smart and intelligent systems. Topics, such as data analytics, decision tree, machine learning, etc., are covered. Additionally, Part 2 features a chapter on supply chain management and blockchain usage in e-waste management.

Part 3 includes emerging green approaches towards a resilient future. It focuses on recycling practises in e-waste industries, along with case studies that add extra flavour.

The book is intended for researchers, academicians, e-waste entrepreneurs, and practitioners. The book will serve as a knowledge base for the researcher community as it encompasses a wide area of intelligent systems as well as sustainable recycling technologies and ICT technologies. The editors would be pleased if the ideas offered in the book were applied to a social cause.

Biswajit Debnath, Abhijit Das,
Potluri Anil Chowdary, and
Siddhartha Bhattacharyya

About the Editors

Abhijit Das received his B.Tech. (IT) from the University of Kalyani, M.Tech. (IT) from the University of Calcutta, and his Ph.D. (engineering) from the Department of CSE Jadavpur University, India.

He has over 18 years of teaching and research experience and more than 50 publications and five edited books of international repute. Presently he is serving as an associate professor in the Department of IT, RCC Institute of Information Technology, Kolkata, India. He had been the head of the Department of IT (Jan 2018–Jan 2020) and has convened various committees at an institutional level.

He has organized and chaired various international conferences and seminars. He has served as a resource person in various institutes and universities and television channels at state and national levels. Currently, six scholars are working with him on different research topics like IoT, e-waste management, data science, quantum computing, object-oriented categorization, etc.

He has published five Indian patents and nine copyrights to date. He serves as a reviewer for many reputed journals and is a professional member of ACM, IEEE, and IETE (Fellow).

He is a professional singer as well and frequently performs on All India Radio Kolkata, DoorDarshan, and various private television channels in India and has more than 30 music albums and singles to his credit.

Biswajit Debnath is a senior research fellow at the Chemical Engineering Department, Jadavpur University. He was a Commonwealth Split-Site Scholar at Aston University, Birmingham, UK (2019–2020). He received his B.Tech and M.E. in chemical engineering in 2013 and 2015, respectively. His area of specialization is waste valorization and sustainability, with a special focus on e-waste and plastic waste. His other research interests include circular economy, climate change, SDGs, supply chain management, sustainable smart city, environmental chemical engineering, etc. He has worked on UKIERI projects and published nearly 80 articles, including conference proceedings and peer-reviewed journals, and contributed to nearly 30 book chapters in his research area published by Springer, Wiley, Elsevier, CRC Press, IEEE, and River Publishers. His h index is 13 with 707 citations. He is a CPCB-certified trainer for the six waste management rules and provided training on e-waste and plastic waste rules on invitation. He has won the best paper award several times at International Conferences. He has delivered 11 invited lectures, including webinars in multiple conferences and seminars. He has been the resource person in five training workshops. Since December 2018, he has been one of the most read authors from his department in ResearchGate. He has completed five collaborative (unfunded) projects with colleagues in the USA, Finland, Saudi Arabia, and India. He is a reviewer of reputed journals, namely *ACS Sustainable Chemistry and Engineering*; *Journal of Material Cycles and Waste Management*; *Journal of Network and Computer Applications; Journal of the Iranian Chemical Society; Environment, Development and Sustainability*; *International Journal of Environmental and Health Research; Journal of Environmental Planning and Management*; *Production & Manufacturing Research*; *Metals*; *Molecules*; *Waste*; *Sustainability*; *Frontiers in Chemistry*; and *Waste Management*. He is also a review editor of *Frontiers in Environmental Engineering in Air Pollution Management; Frontiers in Energy Efficiency Management;* and *Frontiers in High Performance Computing,* published by the Frontiers group. He has edited two books on e-waste management published by CRC Press. He is the co-creator of 'La Sustaina,' which is Kolkata's first upcycled café-cum-store. He also works as a sustainability consultant for national and international clients.

Potluri Anil Chowdary is a managing director of Green Waves Environmental Solutions, the first authorized e-waste collection and handling unit in Andhra Pradesh (unit in Visakhapatnam) since April 2015. He has a graduate diploma in environment resource management from Waiariki Institute of Technology (New Zealand); a master in science (M.Sc) in environmental science branch from GITAM Institute of Science, GITAM University (India); and engineering (chemical technology) from Chaitanya Bharathi Institute of Technology, Osmania University (India). He has worked under the Department of Conservation (Rotorua, New Zealand) on plantation of native flora and pest management in 2014. He has worked on composting and Vermi composting for the Linton Park Community Centre, in Rotorua, under the guidance of Mr. Rick Mansell (Centre Coordinator, Linton Park Community Centre, Rotorua, New Zealand) in 2014. He has also worked for the 'Love Your Water' program organized by Sustainable Coastline on cleaning of fresh and marine water bodies (Rotorua, New Zealand) in 2014. He has received Green Waves Environmental Solutions and has won national awards for excellence in e-waste recycling at an Indian Industry Session (at the 8th Regional 3R Forum in Asia and the Pacific). Yet another golden feather on his crown is the invitation he received from the National Green Tribunal Conference to deliver a talk on e-waste management in Guwahati. On World Environmental Day 2018, he was given the Seva Puraskar award by the Andhra Pradesh Pollution Control Board for his great contribution towards sensitizing the people on e-waste management and for effective recycling of e-waste. "Green Waves Environmental Solutions" was featured as one of the "20 Most Promising E-Waste Management Companies for 2018" in India, recognized by *Silicon India Magazine*. Green Waves Environmental Solutions won the IconSWM Award for its excellence in e-waste recycling at the 8th International Conference on Sustainable Waste Management in 2018.

Dr. Siddhartha Bhattacharyya gained his bachelor's degree in physics, bachelor's degree in optics and optoelectronics, and master's degree in optics and optoelectronics from the University of Calcutta, India, in 1995, 1998, and 2000, respectively. He completed his Ph.D. in computer science and engineering from Jadavpur University, India, in 2008. He is the recipient of the University Gold Medal from the University of Calcutta for his master's. He is the recipient of several coveted awards, including the Distinguished HoD Award and Distinguished Professor Award, conferred by the Computer Society of India, Mumbai Chapter, India, in 2017, the Honorary Doctorate Award (D. Litt.) from the University of South America, and the South East Asian Regional Computing Confederation (SEARCC) International Digital Award ICT Educator of the Year in 2017. He has been appointed as the ACM Distinguished Speaker for the tenure of 2018–2020. He has been inducted into the People of ACM Hall of Fame by ACM, the USA, in 2020. He has been appointed as the IEEE Computer Society Distinguished Visitor for the tenure of 2021–2023. He has been elected as a full foreign member of the Russian Academy of Natural Sciences (RANS) and the Russian Academy of Engineering (REA). He has been elected a full fellow of The Royal Society for Arts, Manufacturers and Commerce (RSA), London, UK.

He is currently serving as the principal of Rajnagar Mahavidyalaya, Rajnagar, Birbhum. He is also serving as the Scientific Advisor of Algebra University College, Zagreb, Croatia. He served as a professor in the Department of Computer Science and Engineering of Christ University, Bangalore. He served as the principal of RCC Institute of Information Technology, Kolkata, India, during 2017–2019. He has also served as a senior research scientist in the faculty of electrical engineering and computer science of VSB Technical University of Ostrava, Czech Republic (2018–2019). Prior to this, he was the professor of information technology at RCC Institute of Information Technology, Kolkata, India. He served as the head of the department from March 2014 to December 2016. Prior to this, he was an associate professor of information technology at RCC Institute of Information Technology, Kolkata, India, 2011–2014. Before that, he served as an assistant professor in computer science and information technology at the University Institute of Technology, The University of Burdwan, India, 2005–2011. He was a lecturer in information technology at Kalyani Government Engineering College, India, during 2001–2005. He is a co-author of six books and the co-editor of 99 books and has more than 400 research publications in international journals and conference proceedings to his credit. He has got two PCTs and 19 patents to his credit. He has been a member of the organizing and technical program committees of several national and international conferences. He is the founding chair of ICCICN 2014, ICRCICN (2015, 2016, 2017, 2018), and ISSIP (2017, 2018) (Kolkata, India). He was the general chair of several international conferences like WCNSSP 2016 (Chiang Mai, Thailand), ICACCP (2017, 2019) (Sikkim, India) and ICICC 2018 (New Delhi, India), and ICICC 2019 (Ostrava, Czech Republic).

He is the associate editor of several reputed journals, including *Applied Soft Computing*, *IEEE Access*, *Evolutionary Intelligence*, and *IET Quantum Communications*. He is the editor of the *International Journal of Pattern Recognition Research* and the founding editor in chief of the *International Journal of Hybrid Intelligence, Inderscience*. He has guest-edited several issues with several international journals. He is serving as the series editor of IGI Global Book Series Advances in Information Quality and Management, De Gruyter Book Series(s) *Frontiers in Computational Intelligence and Intelligent Biomedical Data Analysis (IBDA),* CRC Press Book Series(s) *Computational Intelligence and Applications, Quantum Machine Intelligence and Intelligent Data Driven Technology for Sustainable Environment,* Elsevier Book Series *Hybrid Computational Intelligence for Pattern Analysis and Understanding,* and Springer *Tracts on Human Centered Computing.*

His research interests include hybrid intelligence, pattern recognition, multimedia data processing, social networks, and quantum computing.

He is a life fellow of the Optical Society of India (OSI), India; a life fellow of the International Society of Research and Development (ISRD), UK; a fellow of the Institution of Engineering and Technology (IET), UK; a fellow of Institute of Electronics and Telecommunication Engineers (IETE), India; and a fellow of Institution of Engineers (IEI), India. He is also a senior member of the Institute of Electrical and Electronics Engineers (IEEE), USA; International Institute of Engineering and Technology (IETI), Hong Kong; and the Association for Computing Machinery (ACM), USA.

He is a life member of the Cryptology Research Society of India (CRSI), Computer Society of India (CSI), Indian Society for Technical Education (ISTE), Indian Unit for Pattern Recognition and Artificial Intelligence (IUPRAI), Center for Education Growth and Research (CEGR), Integrated Chambers of Commerce and Industry (ICCI), and Association of Leaders and Industries (ALI). He is a member of the Institution of Engineering and Technology (IET), UK; International Rough Set Society; International Association for Engineers (IAENG), Hong Kong; Computer Science Teachers Association (CSTA), USA; International Association of Academicians, Scholars, Scientists and Engineers (IAASSE), USA; Institute of Doctors Engineers and Scientists (IDES), India; The International Society of Service Innovation Professionals (ISSIP); and The Society of Digital Information and Wireless Communications (SDIWC). He is also a certified Chartered Engineer of the Institution of Engineers (IEI), India. He is on the Board of Directors of the International Institute of Engineering and Technology (IETI), Hong Kong.

Contributors

Chayan Acharya
Department of International Media Studies
Hochschule Bonn-Rhein-Sieg University of Applied Sciences
Germany

Siddhartha Bhattacharyya
Rajnagar Mahavidyalaya
Birbhum, India
and
Algebra University College
Zagreb, Croatia

Sayana Sai Brahmani
Greenwaves Environmental Solutions Private Limited
Vizag, Andhra Pradesh, India

Basabdatta Chakraborty
Techno College Hoogly
West Bengal, India

Avik Chatterjee
Techno College Hoogly
West Bengal, India

Sandip Chatterjee
Electronics Materials & Components
Division (EMCD)
MeitY
Delhi, India

Soumyajit Chatterjee
Chief Technology Officer and Co-founder,
Hoivita Oy
Lappeenranta, Finland

Arup Chattopadhyay
National Institute of Electronics &
Information Technology
Sikkim, India

Potluri Anil Chowdary
Managing Director
Greenwaves Environmental Solutions Ltd.
India

Rohit Roy Chowdhury
Department of Data Science
JIS Institute of Advanced Studies & Research
Kolkata, India

Abhijit Das
Department of Information Technology
RCC Institute of Information Technology
Kolkata, India

Adrija Das
Department of Information Technology
RCC Institute of Information Technology
Kolkata, India

Ankita Das
Department of Data Science and Cyber Security
Institute of Leadership, Entrepreneurship and Development (iLead)
India

Biswajit Debnath
Chemical Engineering Department
Jadavpur University
Kolkata, India
and
Department of Mathematics
ASTUTE, Aston University
Birmingham, UK

Anubha Dubey
Independent researcher and analyst
Acecity-Noida
Uttar Pradesh, India

Tichaona Goto
College of Health Agriculture
and Natural Sciences
Africa University

Arnab Guha
Dublin Business School
Dublin, Ireland

Quinton Chamunorwa Kanhukamwe
Harare Institute of Technology
Zimbabwe Institute of Engineers
Zimbabwe

Preetiman Kaur
Punjab Agricultural University
Ludhiana, Punjab, India

Anthony Phiri
Harare Institute of Technology
Zimbabwe Institute of Engineers
Zimbabwe

Reshma Roychoudhuri
Department of Computer Science
Heritage Institute of Technology
India

Priyankar Roychowdhury
Software Engineer and Independent
Researcher
Ottawa, Canada

Anirbit Sengupta
Department of Electronics and
Communication Engineering
Dr. Sudhir Chandra Sur Institute of
Technology & Sports Complex
India

Shivani Sharma
Punjab Agricultural University
Ludhiana, Punjab, India

Yumnam Jayanta Singh
National Institute of Electronics &
Information Technology
Guwahati, India

Apurva Saxena Verma
Department of Computer Science
Rabindranath Tagore University
Bhopal, M.P., India

Acknowledgments

The editors express their gratitude to CRC Press for publishing the book. We are humbled and grateful to all those who have helped to materialize all the ideas. We sincerely thank the contributing authors and the following persons who helped in bringing the volume to life.

Dr. Gagandeep Singh
Prof. Amar Chandra Das
Ms. Ankita Das
Mrs. Ratna Das
Mr. Abhijit Debnath
Ms. Abhipsa Das
Mr. Amritesh Das
Dr. Divis Murali
Dr. Kameswar Rao
Mr. Pranesh Varma
Late Manindra Debnath
Late Guru Ramkrishna Goswami
Team Green Waves

1 Introduction

Biswajit Debnath, Anil Potluri, Abhijit Das, and Siddhartha Bhattacharyya

1.1 POST COVID-19 SITUATION OF E-WASTE MANAGEMENT

E-waste is globally a big challenge as the rate of outdated products increases with the development of technologies and innovations in the electronics sector. Since, the inception of computers, the global consumer electronic market has increased exponentially to become a multibillion-dollar industry (Debnath 2020). In 2019, global e-waste generation reached 53.6 million metric tonnes (Mt) from 44.7 Mt in 2016 (Baldé et al. 2017; Forti et al. 2020). The same report published by United Nations University in association with International Telecommunication Union (ITU) & International Solid Waste Association (ISWA) predicts that by 2030, discarded electronic products will reach 74 Mt, which is an alarming number (Forti et al. 2020). However, the year 2020 will remain a milestone for humankind due to the outbreak of the COVID-19 pandemic (Goldstein 2020). During the pandemic, it was predicted that there would be a considerable increase in electronic goods consumption and its subsequent discordance. However, in the first two-quarters of the pandemic, a reduction of nearly 4.9 Mt of e-waste generation in the future e-waste stream was recorded (Baldé and Kuehr 2021). The pandemic also recorded a significant lowering in e-waste collection, not only in low- and middle-income countries but also in wealthy countries (Baldé and Kuehr 2021).

Post-COVID-19, the whole world has started rethinking its waste management strategies. While it seems to be good news that e-waste generation reduced due to the pandemic, several other issues have emerged. E-waste recycling companies were affected in material procurement, especially e-waste related to information technology (IT) software companies as they were operating in a 'work from home' format. In this format, companies didn't show the interest in disposing of e-waste on certain timelines. Many small- and medium-scale industries were shut down during this phase. Due to shortage of material flow as well as demand end, metal rates were sinking, which affected the sales of recovered metals from e-waste. Many e-waste dismantling companies with small capacities incurred huge losses due to an increase in work capital. In addition to that, there was reduction in e-waste flow in the market as lots of consumers were busy recovering business rather than focusing on waste management. For example, the educational institutes and a few MNCs were more concerned about recovering business and maintaining brand image. It was also observed that a lot of biomedical devices were discarded from hospitals in the post-Covid timeline.

DOI: 10.1201/9781003317050-1

1.2 TECHNOLOGICAL SYSTEMS FOR E-WASTE MANAGEMENT

E-waste is a very heterogeneous material, and it contains an assortment of materials, including metals, polymers, and siliceous materials, including glass. Such discarded e-waste streams thus trap a huge amount of metallic and non-metallic resources that relegates urban mining of e-waste as not an option anymore, but rather a necessity. The term 'urban mining' is synonymous with resource recovery from e-waste. There are several technologies developed for management of e-waste. These technologies can be divided into two categories: (a) e-waste recycling and valorization technologies and (b) e-waste supervision technologies. E-waste recycling and valorization technologies are primarily wet processes that are carried out in the recycling unit itself. There are several technologies for e-waste recycling and valorization, such as mechanical recycling, hydro-metallurgical, pyro-metallurgical, and bio-metallurgical technologies (Alam et al. 2022). In the past two decades, these technologies have become conventional technologies except bio-metallurgical technologies (Debnath et al. 2018). There are new and evolving recycling technologies that aim more toward e-waste valorization than just recovery. For example, green pyrolysis is aimed toward reducing environmental impact and at the same time recovering as much resource as possible (Andooz et al. 2021). Use of green solvents is also evolving, which has the potential to make hydro-metallurgical processes greener. Green catalysts are being developed from different parts of e-waste, including printed circuit boards, waste toner, Li-ion batteries, and e-waste plastics (Rodríguez-Padrón et al. 2020). This valorization route can open new doors. Overall, the technological advancement in recycling and valorization of e-waste will lead towards a sustainable system.

However, this array of recycling technologies cannot save the e-waste recycling industry alone. This is where the e-waste supervision technologies come into the picture. These are 'dry-lab' sort of technologies that can be carried out or monitored ex-situ. It is equally important to have a strong and efficient supply chain as it is to have a sustainable recycling system. Technological advancement in the field of Information and communication technology (ICT) is showing the path towards developing an efficient supply chain network (SCN). The major factor here is e-waste collection and transportation. E-waste SCN can be really complex, which makes things difficult. State-of-the-art literature shows that the Internet of Things (IoT) has been utilized to streamline e-waste collection and monitor the transport pathways using sensors (Razip et al. 2022). There are a number of instances where intelligent algorithms are being deployed to optimize the transportation routes as well as overall collection efficiency. Intelligent tools such as game theory, machine learning, MCDM techniques, fuzzy mathematics, etc. are being used for supply chain optimization (Doan et al. 2019). Classical methods such as agent-based modelling and principal component analysis (PCA), to evolving methods such as artificial neural network (ANN) and advanced system dynamics modelling, are enriching the area of e-waste supply chain (Ghalehkhondabi and Ardjmand 2020; Debnath 2022; Debnath et al. 2022). Blockchain technology, with its unique characteristics of traceability, is the new buzz in the supply chain world. With the feature of traceability, e-waste supply chain monitoring will become much easier (Khan and Ahmad 2022). It will be the partial truth to state that these digital technologies can just be e-waste supervision technologies. Their reach is more than that. Cloud-based platforms have been used for the development of remanufacturing e-waste. Blockchain can improve the monitoring of the whole lifecycle of e-waste, which will play a pivotal role towards achieving a circular economy. In a nutshell, these ICTs, combined with greener recycling technologies, will ensure the development of intelligent, smart, and sustainable systems, individually or collectively.

1.3 TOWARDS SMART AND SUSTAINABLE E-WASTE MANAGEMENT SYSTEMS

The e-waste management system is complex, and so is its SCN, which plays a key role in ensuring business sustainability. There are several issues and challenges in the e-waste management system. As mentioned, technologies such as IoT, cloud computing, and blockchain are paving the path towards resilient e-waste supply chain management. Despite the odds, development in IT plays a pivotal role in developing intelligent systems that can ensure a sustainable and circular e-waste management system. A smart, intelligent, and sustainable e-waste management system is essential for a better future. These achievements can be well attributed to the proliferation of smart and intelligent systems. In addition to that, emerging green technologies need to be explored, and more pilot-scale developments are the key here. Hence, an optimal combination of new technologies and existing ones is needed to solve the e-waste problem and shift towards truly sustainable smart systems.

REFERENCES

Alam, Tanvir, Rabeeh Golmohammadzadeh, Fariborz Faraji, and M. Shahabuddin. "E-waste recycling technologies: an overview, challenges and future perspectives." *Paradigm Shift in E-waste Management* (2022): 143–176.

Andooz, Amirhossein, Mohammad Eqbalpour, Elaheh Kowsari, Seeram Ramakrishna, and Zahra Ansari Cheshmeh. "A comprehensive review on pyrolysis of e-waste and its sustainability." *Journal of Cleaner Production* (2021): 130191.

Baldé, C. P., and R. Kuehr. *Impact of the COVID-19 Pandemic on E-waste in the First Three Quarters of 2020*. United Nations University (UNU)/United Nations Institute for Training and Research (UNITAR) – co-hosting the SCYCLE Programme, Bonn (Germany), 2021.

Baldé, Cornelis P., Vanessa Forti, Vanessa Gray, Ruediger Kuehr, and Paul Stegmann. *The Global E-waste Monitor 2017: Quantities, Flows and Resources*. United Nations University, International Telecommunication Union, and International Solid Waste Association, 2017.

Debnath, Biswajit, Chowdhury, Ranjana, & Ghosh, Sadhan Kumar (2018). Sustainability of metal recovery from E-waste. *Frontiers of Environmental Science & Engineering* 12, 1–12. 10.1007/s11783-018-1044-9.

Debnath, Biswajit. "Towards Sustainable E-Waste Management through Industrial Symbiosis: A Supply Chain Perspective." In *Industrial Symbiosis for the Circular Economy*, pp. 87–102. Springer, Cham, 2020.

Debnath, Biswajit, Rihab El-Hassani, Amit K. Chattopadhyay, T. Krishna Kumar, Sadhan K. Ghosh, and Rahul Baidya. "Time evolution of a supply chain network: kinetic modeling." *Physica A: Statistical Mechanics and Its Applications* 607 (2022): 128085.

Debnath, Biswajit, Das, Ankita, and Das, Abhijit (2022). Towards circular economy in e-waste management in India: Issues, challenges, and solutions, *Circular Economy and Sustainability* (pp. 523–543). 10.1016/b978-0-12-821664-4.00003-0.

Doan, Linh Thi Truc, Yousef Amer, Sang-Heon Lee, Phan Nguyen Ky Phuc, and Luu Quoc Dat. "E-waste reverse supply chain: A review and future perspectives." *Applied Sciences* 9, no. 23 (2019): 5195.

Forti, Vanessa, Cornelis P. Balde, Ruediger Kuehr, and Garam Bel. *The Global E-waste Monitor 2020: Quantities, Flows and the Circular Economy Potential.* United Nations University (UNU)/United Nations Institute for Training and Research (UNITAR) – co-hosted SCYCLE Programme, International Telecommunication Union (ITU) & International Solid Waste Association (ISWA), Bonn/Geneva/Rotterdam, 2020.

Ghalehkhondabi, Iman, and Ehsan Ardjmand. "Sustainable e-waste supply chain management with price/sustainability-sensitive demand and government intervention." *Journal of Material Cycles and Waste Management* 22, no. 2 (2020): 556–577.

Goldstein, Joseph L. "The Spanish 1918 Flu and the COVID-19 disease: the art of remembering and foreshadowing pandemics." *Cell* 183, no. 2 (2020): 285–289.

Khan, Atta Ur Rehman, and Raja Wasim Ahmad. "A blockchain-based IoT-enabled e-waste tracking and tracing system for smart cities." *IEEE Access* 10 (2022): 86256–86269.

Razip, Marym Mohamad, K. S. Savita, Khairul Shafee Kalid, Mohammad Nazir Ahmad, Maryam Zaffar, Eidia Erriany Abdul Rahim, Dumitru Baleanu, and Ali Ahmadian. "The development of sustainable IoT E-waste management guideline for households." *Chemosphere* 303 (2022): 134767.

Rodríguez-Padrón, Daily, Zeid A. ALOthman, Sameh M. Osman, and Rafael Luque. "Recycling electronic waste: prospects in green catalysts design." *Current Opinion in Green and Sustainable Chemistry* 25 (2020): 100357.

Part 1

Sustainable Technologies

2 Application of New Information Technologies for ICT Waste Management

Anirbit Sengupta, Adrija Das, Basabdatta Chakraborty, Avik Chatterjee, Biswajit Debnath, Ankita Das, Reshma Roychoudhuri, Priyankar Roychowdhury, and Abhijit Das

2.1 INTRODUCTION

Waste electrical and electronic equipment (WEEE) is also known as e-waste or electronic waste (Debnath et al. 2016). Information and communication technology waste or ICT waste has the alias of e-waste (Emmanouil et al. 2013). Electronic gadgets are meant to make our lives happier and simpler, but recycling and disposing of the toxic substances of electronic equipment are a health nightmare (Wath et al. 2010). Compared to traditional municipal wastes, e-waste contains circuit boards that frequently include hazardous compounds such as lead, mercury, and cadmium, as well as nickel, beryllium, and zinc. These substances can cause threats to human health as well as to the global environment (Widemar et al. 2005). The hazardous materials lead to long-term diseases of the nervous system, kidney and bones, reproductive systems, and endocrine systems as a few of them are carcinogenic in nature. Chlorofluorocarbon (CFC) and hydrochlorofluorocarbon (HCFC) are responsible for the Greenhouse Effect, and these gases are emitted from extraction and refinement of primary raw materials of e-waste (Pan et al. 2022).

E-waste generation is increasing rapidly and is the fastest growing waste stream in the world (Goosey et al. 2009; Hossain et al. 2015). This trend of e-waste generation is shown in Figure 2.1. In 2019, global generation of e-waste was 53.6 million metric tons (Mt), which is equivalent to 5,516 Eiffel towers (Forti et al. 2020; Debnath et al. 2022). Hence, proper management of e-waste is imperative. Proper management of ICT waste can help to build a sustainable tomorrow. Waste reduction in industries can be done by adopting (a) inventory management, (b) production process management, (c) volume reduction, (d) recovery and reuse, and (e) recycling (Rao 2014). E-waste recycling is emerging as one of the important approaches towards resource circulation and urban mining (Das et al. 2020).

DOI: 10.1201/9781003317050-3

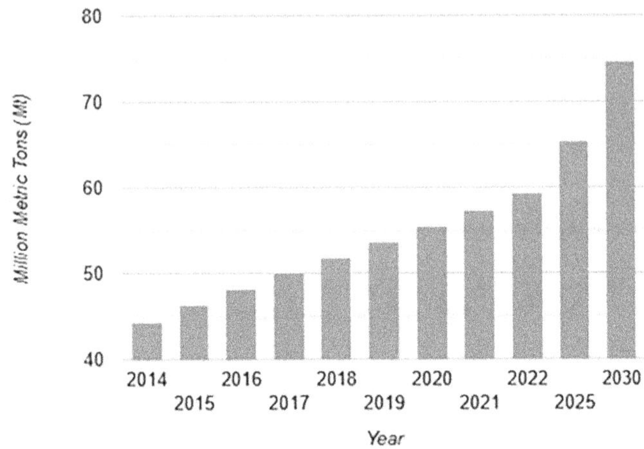

FIGURE 2.1 E-waste generation worldwide (Yearwise). (Adapted from theroundup.org)

Not only does it reduce global pollution, but it also saves natural resources (energy), conserving landfill space and cutting down on production waste (Rao 2014). Recycling industries are offering various kinds of jobs that can boost the economy of any country in large aspects (Patwa et al. 2021).

The most recent paradigm shift is the fourth industrial revolution, or 'Industry 4.0.' This is made possible by emerging information technologies such as Internet of Things (IoT), cloud computing, edge computing, Big Data analytics, blockchain technologies, etc (Gupta and Bedi 2018;Surapaneni et al. 2018; Sahoo et al. 2020; Das et al. 2020). ICT can play an important role in the management of ICT waste (Debnath et al. 2022). In Malaysia, a mobile application was introduced for public end-users to dispose of their household waste (Kang et al. 2020). IoT-enabled models were found to manage e-waste in smart cities (Surapaneni et al. 2018). Blockchain technologies can be used to trace e-waste (Dindarian et al. 2019). An innovative model was built using blockchain technology to join the Grid and information exchange and purchase energy between nodes for smarter cities, which can help to manage wastes as well (Pieroni et al. 2018). Chen and Ogunseitan (2021) have highlighted the opportunities of using blockchain towards achieving the concept of zero e-waste. Recently, Sahoo and Haldar (2021) developed a blockchain-based prototype using Ethereum for sustainable e-waste management. Khan and Ahmad (2022) have also developed a blockchain-based, IoT-enabled ICT waste tracking system that is tested on Ethereum.

This body of work indicates that new and emerging information technology is already being applied for sustainable management of ICT waste. Hence, these technologies can play a key role in monitoring and managing ICT waste supply chains. Despite the fact that there is a great deal of software, artificial intelligence (AI), and coding involved, these technologies also serve as the predecessors of future ICT waste because they require communication hardware components in order to operate and communicate. The idea of 'poison-kills-poison' is welcomed here since the aforementioned technologies can be very useful for managing ICT waste. In this chapter, we present a projection of the possible applications of the new and emerging information technologies in ICT waste management. In addition, industrial and sustainability aspects are also explored with a proposed framework for sustainable ICT waste management.

2.2 METHODOLOGY

The majority of the chapter is based on already published literature, followed by further analysis. An extensive literature review was done to acquire information on new information technology for ICT waste management. The literature review used a number of single and multiple keywords, including 'ICT waste,' 'e-waste,' 'e-waste + cloud computing,' 'e-waste + green computing,' 'e-waste + IoT,' 'e-waste + blockchain,' and 'ICT waste + sustainability.' The gathered documents were categorized and sorted. Where appropriate, the cross-references were also assessed. To get data regarding the chapter's composition, more analysis was conducted. Last but not least, a brainstorming session was held to develop a sustainable framework for ICT waste management using new information technology.

2.3 APPLICATION OF NEW INFORMATION TECHNOLOGIES FOR ICT WASTE MANAGEMENT

2.3.1 Internet of Things (IoT)

As new technologies are adding tons of e-waste to the world, they are also inventing measures of disposal of such waste. One such technology is IoT (Bandyopadhyay and Sen 2011). A group of researchers from Monash University in Malaysia (Kang et al. 2020) have explained an example of a smart e-waste collection system using IoT. Their research describes an efficient e-waste management system that claims to significantly recycle household electronic disposals in Malaysia, as Southeast Asian countries lead in e-waste generation in the world. Their proposed system consists of one mobile application to guide the user to a proper disposal place, microcontroller-equipped smart e-waste collection boxes with ultrasonic sensing and WiFi capability, and cloud database servers. These systems are capable of collecting e-waste and processing it for reuse, maintaining sustainability. A solution to disposal and proper management of solid waste by using IoT is described in another case study (Surapaneni et al. 2018). Their proposed model is called Integrated Solid Waste Management (ISWM) and is based on an Indian context. ISWM works on the policy of: (a) minimizing usage, (b) promoting the use of refill containers, (c) introducing incentives for customers, (d) encouraging environmentally friendly design of products, and (e) promoting the development of eco-industrial plants. ISWM integrates and incorporates an ultrasonic and motion detection sensor, GPS, GPRS and RFID, and a moisture sensor. Use of bio-organic species such as aloe vera in place of conventional electronic materials can be a sustainable way of minimizing the generation of e-waste (Lim et al. 2017). Real-time detection of trash disposal breach with means of a serverless IoT-enabled framework is possible due to effective edge computing architectures like recycle.io (Al-Masri et al. 2018).

The overall architecture of recycyle.io integrates edge computing competencies with waste management approaches to overcome the challenges of urban waste management. Recent trends of usage of electronic products show that the current focus is on circular economy, which opens up a golden opportunity to move towards a resource-resilient society (Bangs et al. 2016; Reuter and Van Schaik 2016; Ramadoss et al. 2018). Regarding the issue of data leakage and security risk of recycling and reusing electronic devices, an effective mechanism was discussed in detail (Schafer 2014). RFID is a strong method in waste management that can also simultaneously be applied in e-waste management (Tao and Xiang 2010; Ramesh et al. 2017). The sensor-fitted high-powered drone is operated remotely. The drone is equipped with GSM and GPS modules and can sense various types of e-hazardous waste. These drones traverse across the dump yards where all the waste is dumped and locate the high-risk waste via their sensors. This information is then sent to the central server, where it is stored using timestamps. The workers are informed through HMD using augmented reality, which helps the workers to locate highly toxic places. At regular intervals, the places will be sanitized. An indoor localization system, LANDMARC, was developed with the intention of improving the utilization of electronic devices based on the Disk Sensing Model and the Exponential Sensing Model to ensure proper coverage, detection, and location (Ullah and Sarkar 2018). Also, IoT-based wireless sensor networks that use Raspberry Pi technology can be developed following the model described by Lata and Singh (2016).

2.3.2 Cloud Technology

Cloud computing can have a vital role to control and manage e-waste through various ways. Zhang et al. (2010) proposed a cloud-based information platform, to improve the efficiency of information flow management and the e-waste recycling system considering a take-back system of e-waste and a multiple stakeholder recycling network. The information platform includes three sub-clouds, namely collection platform, recycling platform, and transaction platform, for materials recovered. Recyclers have the option of leasing a SaaS-based ERP system through the proposed recycling platform. Wang and Wang (2015) introduced the cloud manufacturing paradigm in the e-waste remanufacturing environment. To facilitate e-waste remanufacturing, they developed a revolutionary cloud-based system, WRCloud. The e-waste recycling capabilities are provided in the cloud as adaptable services. WRCloud is a three-layer architecture: user layer, cloud service coordinator layer, and remanufacturing layer. This also includes a QR code-based e-waste tracking system. The model is validated with multiple case implementation as well. However, they argue that since both manufacturers and recyclers must put in additional configuration work before using the cloud, early adoption of the cloud idea in the remanufacturing industry may be challenging (Wang and Wang 2015; Wang et al. 2014). Jayarajan et al. (2018) developed a smart cloud-enabled system for monitoring e-waste dumpsites using a smartphone application. Recently, Das et al. (2020) proposed a cloud-based framework for e-waste inventorisation. They have focused on record-keeping of sales data of electronic products from the e-commerce sector as well as offline sales. The proposed system is a three-layer system that aims to trace online and offline sales from Goods and Service Tax (GST) data. Despite multiple benefits and befitting applications, there are many concerns regarding data security and reverse engineering. It is expected that integration of two or more technologies with cloud could be a potential way out.

2.3.3 Blockchain Technology

Blockchain technology is one of the major technologies that many countries and organizations are following to control WEEE. According to Casino et al. (2019), blockchain is the technique that permits transactions to be validated by an untrustworthy set of participants. The blockchain system organizes information into a chain of blocks, with each block storing a collection of Bitcoin transactions that occurred at a specific time. A chain is formed when blocks are linked together by a reference to the previous block. Different applications are being developed in order to reduce administration and audit costs and increase transparency and efficiency in the certification and verification of documents by using blockchain. The integration between blockchain and the IoT was intended to address the previous drawbacks in addition to maintaining reliable data (Diaz et al. 2016).

Schluep (2014) has shown that the e-waste management supply chain is one of the most complex supply chains as multiple stakeholders are involved in every phase of it. In terms of digitalization, the waste management industry is far behind other sectors and relies mainly on the manual record-keeping process in most cases (Deloitte 2014).

The blockchain platform ChemChain is a project developed by Chemycal, and according to the company's website (ChemChain 2018), ChemChain is designed to facilitate the transfer and tracking of information on hazardous chemicals. It helps every stakeholder to be compliant with regulations, such as, for example, REACH (the Chinese Due Diligence Guidelines for Responsible Mineral Supply Chain, the EU Conflict Minerals Regulation, and Registration, Evaluation, Authorisation and Restriction of Chemicals) in Europe (CCCMC 2015) and Proposition 65 in California (ChemChain 2018). According to Dindarian (2019) every time a raw material or product changes hands, from manufacture to sale, information on chemicals continues to flow and can be documented. It dramatically reduces the complexity and the costs of the overall mechanism of handling e-waste. In addition, it also increases awareness among consumers about the chemicals used in products and their risks. Netherlands-based company Bundel uses 'product as a service' for household white goods as their business model; it has partnered with nine other organizations, including Circle Economy and Rabobank, to deploy blockchain-enabled technology within their circular pay-per-use business models, in which it is planned to use smart contracts and virtual currencies (Circle-Economy 2018). Kumar et al. (2017) proposed a new business model where reliable and transparent information on type, quantity, and quality of waste can be collected in real time, helping recyclers to receive waste through legal channels and encouraging users to recycle end-of-life products. In the recent past, a tokenised recycling process based on blockchain technology has been proposed. The tokens have been used for paying recyclers with digital currency, while providing customers with the opportunity to trace their waste from the moment it goes into the bin until it enters a recycling facility (Sensa Networks 2019). Swachhcoin is a project that combines AI, Big Data, IoT, smart contract, and blockchain in a platform. It facilitates its users with incentives in terms of tokens to collect and recycle waste (Swachhcoin 2017). The company has used an open-source blockchain-enabled technology to decentralize waste management, increase transparency, and eliminate middlemen within the supply chain. In Recereum, the principal aim is to reduce the cost of waste sorting and collection. It is an ecosystem for individual users who earn tokens while separating and recycling their waste at origin (Recereum 2018). Blockchain technology is evolving in a fast and growing environment. Concerns over the scalability of public blockchains have led to efforts to improve the consensus algorithms used by blockchains. Ethereum, for example, is planning to replace its Proof of Work (PoW) algorithm with a much more scalable Proof of Stake (PoS). Newer blockchain platforms are using considerably more optimized platforms to begin with, in order to avoid scalability issues in the future (Dindarian and Chakravarthy 2019).

Given the complexity of the waste management industry, it must be emphasized that the scalability of these projects depends on the level of involvement of all stakeholders like miners, smelters, manufacturers, government agencies, customers, recyclers, etc., and for them to come together for this emerging technology to reach its full potential (Kshetri 2018).

2.3.4 Green Computing

Green computing, also called green technology, is the environmentally responsible use of computers and related resources. One of the first instances of trying to go green is probably the Energy Star program started in 1992. Under its aegis, computers were marked energy efficient if they minimized carbon emission and maximized efficiency. As organizations started becoming green aware, they started promoting green practices among their employees, like switching off their monitors when going on breaks or at the end of the day. Non-server machines or machines not running batch programs were switched off at the end of the day. Overdue course manufacturers also started becoming aware of green practices, and as an early manifestation of that, the monitors started coming with a 'sleep mode.' We have come a long way since then, and there are more holistic approaches to green computing now, and green practices have influenced every stage in the life cycle of a computer. Green use has become the call of the day, and this refers to reducing energy consumption and emission of computers by using them in an energy-efficient manner. Let's consider data centres as an example. 'Data centre servers use 50 times the energy per square foot as an office [does],' says Mark Bramfitt, principal program manager at PG&E. Data centres are the main reason behind energy consumption. Energy consumed by data centres in the United States and worldwide doubled from 2000 to 2005, according to Jonathan Koomey, a consulting professor at Stanford University and staff scientist at Lawrence Berkeley National Lab. As a solution, real-time web-based multi-tier interactive e-green is proposed that educates computer users about managing computers and their accessories in an innovative way (Okewu et al. 2017). The impact of green computing is quite vast. A study showed green computing effects in socio-economic and environmental aspects, especially those of e-waste. It explains key barriers of global adaptation of green computing (Airehrour et al. 2019). Green computing became an interesting domain of research nowadays. Detailed analysis of current trends and challenges of green computing is discussed in current research (Patil et al. 2018). For achieving a sustainable environment, everyone in the society should adopt green technologies. Students are the budding citizens for any country; how students are practising green computing is also a concern that has been explored by a study (The Cognizance of Green Computing Concept and Practices among Secondary School Students: A Preliminary Study). Finally, green disposal ensures that once rejected, computer subsystems may either be recycled or end up as e-waste. Green disposal refers to energy-efficient recycling of the computer subsystem for reuse. It also refers to using proper technologies to recycle e-waste and thus extract precious metals in the process. It has been shown that numerous aspects of green computing are possible for IoT computing, analyzing critical concepts, challenges, and remediation (Muniswamaiah et al. 2020). Heuristic problems related to e-waste are effectively solved by a highly efficient computational technique that plays an important role of e-waste segregation as well. Scope of artificial intelligence in management of e-waste is also described in the same study (Agarwal et al. 2020). In a unique study, a smart dustbin was designed that was eventually connected to green computing. This alternatively reduced pollution and promoted creating more recyclable products (Ramanujam et al. 2020). Virtualization can be used as an optimized technique for resource utilization, load balancing, e-waste collection, and recycling (Anwar et al. 2018). The manufacturing of computers with non-hazardous particles has been described in an academic study that shows a way to more sustainable green products (Sengupta 2019).

2.4 FINDINGS AND ANALYSIS

2.4.1 Industrial Perspective of New IT Application for ICT Waste Management

Data science and machine learning can play an important role in the e-waste management industry like countless other industries (Wang et al. 2021). However, to obtain the benefit, the primary requirement is datasets about e-waste based on which algorithms and mechanisms of data analytics work (Das et al. 2020). The dataset formation about e-waste can start from the manufacturing of an electronic gadget or appliance itself. Every manufacturer is aware of and can maintain a database of possible e-waste components available from a product it has built. When their products are sold either by e-commerce websites or in-store retailers, they are informed that their products are out to the consumers. The actual e-waste data tracking starts from the point when a consumer stops using a product (Debnath et al. 2016). Currently, most of the gadgets end up staying in a drawer in one's lifetime or dumped into the recyclable or e-waste bins provided by the local government responsible for trash collection or e-waste recycling facilities provided by certain stores such as 'BestBuy' (Debnath et al. 2020). However, changes can be made in the best interest of the e-waste industry for obtaining data and the product itself from a consumer directly by providing consumers coupons or store discounts when a gadget is returned to an e-waste facility and reported by a consumer. The facility on further inspection can mark in the manufacturer database the status of each component whether it can be reused or treated as waste. Then, actions can be taken according to the components (Debnath et al. 2016). Data analysis can be performed on the data, which can provide insight into which components of which product of which manufacturer are going where. Machine learning can predict what can be the situation or trend in a particular location in a particular time period based on the sales of gadgets as well as waste datasets (Rai et al. 2021). The information about the current trends and future predictions can be useful for decision making in the e-waste industry and its stakeholders.

2.4.2 Solutions for Sustainable ICT Waste Management

ICT waste management is a big concern worldwide, specifically with the bulk consumers where technological evolution dictates the product desuetude. As discussed above, the new information technologies have enough capabilities for management of e-waste. However, their efficiency lies mostly in monitoring the management system rather than recovering resources and recycling itself. Based on the detailed literature and the author's experience in the field of e-waste management, the following solutions are provided, which will be suitable, especially for the bulk consumers:

i. A proper ICT waste management policy should be developed in the respective organizations for proper handling, disposal, and monitoring of ICT waste. Some organizations do have sustainability policies in place; amendments are suggested for streamlining ICT waste management practices.
ii. Most of the IT companies and other MNCs, who are the bulk consumers as well as bulk e-waste generators, are more into the practise of ESG, i.e., Environment, Social and Governance, which is often taken care of by a dedicated manager or a sustainability manager. It is suggested to provide them proper training about ICT waste management and bring it into a well-established process.
iii. ISO-based frameworks such as ISO 9000, ISO 18000, and ISO 27000 can be implemented for better inventorization of ICT waste, for better environmentally friendly decision making, and most of all for better information security systems, respectively, in an organization (Roychoudhuri and Debnath 2020; Debnath et al. 2020).
iv. It is suggested that researchers working on e-waste management should look for developing circular designs for better remanufacturing of e-waste (Wang and Wang 2015). In addition to that, solutions should be sought for utilizing both chemical technologies and new information technologies for sustainable e-waste management.

2.5 CONCLUSION

ICT waste, aka e-waste, is one of the fastest growing streams of the world. With the aid of chemical technologies, it is possible to recover metals and other value-added materials from e-waste. But the challenge remains with better channelization and management of e-waste. In this chapter, the potential of new and emerging information technology for the sustainable management of e-waste was explored. The findings suggest that technologies such as cloud computing, edge computing, and IoT can be utilized for streamlining e-waste management, especially in monitoring the supply chain. On the other hand, blockchain technology was found to be useful for supply chain transparency. It was recommended that a combination of two or more new IT should be explored for better solutions. In addition to that, an amalgamation of chemical and new information technologies should be explored for sustainable management of ICT waste. It is expected that more studies on this line will create a truly inter-disciplinary area of research focused on waste management.

ACKNOWLEDGEMENTS

The authors acknowledge the support received from their respective institutes. Any opinions, findings, and conclusions expressed in this material are those of the author (s) and do not necessarily reflect the views of the affiliated universities of the authors.

REFERENCES

Agarwal, Vernika, Shivam Goyal, and Sanskriti Goel. "Artificial intelligence in waste electronic and electrical equipment treatment: Opportunities and challenges." In *2020 International Conference on Intelligent Engineering and Management (ICIEM)*, pp. 526–529. IEEE, 2020.

Airehrour, David, Marianne Cherrington, Samaneh Madanian, and Jagjeet Singh. "Reducing ICT carbon footprints through adoption of green computing." In *Academy of Economic Studies in Bucharest. Department of Economic Informatics and Cybernetics*, 2019. 10.12948/ie2019. 04.17

Al-Masri, Eyhab, Ibrahim Diabate, Richa Jain, Ming Hoi Lam, and Swetha Reddy Nathala. "Recycle. io: An IoT-enabled framework for urban waste management." In *2018 IEEE international conference on big data (big data)*, pp. 5285–5287. IEEE, 2018.

Anwar, Sidra, Mahjabeen Ghaffar, Fatima Razzaq, and Bushra Bibi. "E-waste reduction via virtualization in green computing." *American Scientific Research Journal for Engineering, Technology, and Sciences (ASRJETS)* 41, no. 1 (2018): 1–11.

Bandyopadhyay, Debasis, and Jaydip Sen. "Internet of things: Applications and challenges in technology and standardization." *Wireless Personal Communications* 58, no. 1 (2011): 49–69.

Bangs, Colton, Christina Meskers, Thierry Van Kerckhoven, and Umicore Precious Metals Refining. "Trends in electronic products—the canary in the urban mine." *Proceedings of the Electronic Goes Green* (2016): 7–9.

Casino, Fran, Thomas K. Dasaklis, and Constantinos Patsakis. "A systematic literature review of blockchain-based applications: Current status, classification and open issues." *Telematics and informatics* 36 (2019): 55–81.

CCCMC. "Chinese Due Diligence Guidelines for Responsible Mineral Supply Chains." Project Brief (2015).

ChemChain. (2018). Available from: https://www.chemcha.in/ Visited: 13/06/2019

Chen, Mengjun, and Oladele A. Ogunseitan. "Zero e-waste: Regulatory impediments and blockchain imperatives." *Frontiers of Environmental Science & Engineering* 15, no. 6 (2021): 1–10.

Circle-Economy. How Blockchain can advance the circular economy, 2018. Available from: https://www.circle-economy.com/

Das, Ankita, Biswajit Debnath, Nipu Modak, Abhijit Das, and Debasish De. "E-waste inventorisation for sustainable smart cities in India: A cloud-based framework." In *2020 IEEE International Women in Engineering (WIE) Conference on Electrical and Computer Engineering (WIECON-ECE)*, pp. 332–335. IEEE, 2020.

Debnath, Biswajit, Roychoudhuri, Reshma, & Ghosh, Sadhan K. (2016). E-Waste Management – A Potential Route to Green Computing. Procedia Environmental Sciences, 35, 669–67510.1016/j.proenv.2016.07.063.

Debnath, Biswajit, Priyankar Roychowdhury, and Rayan Kundu. "Electronic components (EC) reuse and recycling–a new approach towards WEEE management." *Procedia Environmental Sciences* 35 (2016): 656–668.

Debnath, Biswajit, Jaafar M. Alghazo, Ghanzafar Latif, Reshma Roychoudhuri, and Sadhan Kumar Ghosh. "An analysis of data security and potential threat from IT assets for middle card players, institutions and individuals." In *Sustainable Waste Management: Policies and Case Studies*, pp. 403–419. Springer, Singapore, 2020.

Debnath, Biswajit, Ankita Das, and Abhijit Das. "Towards circular economy in e-waste management in India: Issues, challenges, and solutions." In *Circular Economy and Sustainability*, pp. 523–543. Elsevier, 2022.

Debnath, Biswajit, Anil Potluri, and Abhijit Das. "Introduction." In B. Siddhartha et al. (eds.) *Paradigm Shift in E-waste Management: Visions for the Future*, pp. 1–6. CRC PRESS, UK, 2022.

Deloitte, B. I. O. "Development of guidance on extended producer responsibility (EPR): Final report." *Retrieved September* 5 (2014): 2014.

Dindarian, Azadeh. Overview of ChemChain, Personal communication with L. Zullo CEO of Chemycal, 2019.

Dindarian, Azadeh, and Sid Chakravarthy. "Traceability of electronic waste using blockchain technology." *Issues in Environmental Science and Technology* (2019): 188–212.

Díaz, Manuel, Cristian Martín, and Bartolomé Rubio. "State-of-the-art, challenges, and open issues in the integration of Internet of things and cloud computing." *Journal of Network and Computer Applications* 67 (2016): 99–117.

Emmanouil, Maria-Chrysovalantou, Emmanouil Stiakakis, Maria Vlachopoulou, and Vasiliki Manthou. "An analysis of waste and information flows in an ICT waste management system." *Procedia Technology* 8 (2013): 157–164.

Forti, V., Baldé, C. P., Kuehr, R., & Bel, G. (2020). The global e-waste monitor 2020. *United Nations University (UNU), International Telecommunication Union (ITU) & International Solid Waste Association (ISWA), Bonn/Geneva/Rotterdam.* 120.

Goosey, Martin, Gary Stevens, and Henryk Herman. Electronic Waste Management. Vol. 27. Royal Society of Chemistry, 2009.

Gupta, Neha, and Punam Bedi. "E-waste management using blockchain based smart contracts." In *2018 International Conference on Advances in Computing, Communications and Informatics (ICACCI)*, pp. 915–921. IEEE, 2018.

Hossain, Md Sahadat, Sulala MZF Al-Hamadani, and Md Toufiqur Rahman. "E-waste: A challenge for sustainable development." *Journal of Health and Pollution* 5, no. 9 (2015): 3–11.

Jayarajan, P., S. Thenmozhi, R. Maheswar, S. Malathy, and R. Udaiyakumar. "Smart cloud enabled e-waste management system." In *2018 International Conference on Computer Communication and Informatics (ICCCI)*, pp. 1–5. IEEE, 2018.

Kang, Kai Dean, Harnyi Kang, I. M. S. K. Ilankoon, and Chun Yong Chong. "Electronic waste collection systems using Internet of Things (IoT): Household electronic waste management in Malaysia." *Journal of Cleaner Production* 252 (2020): 119801.

Khan, Atta Ur Rehman, & Ahmad, Raja Wasim (2022). A Blockchain-Based IoT-Enabled E-Waste Tracking and Tracing System for Smart Cities. IEEE Access, 10, 86256–86269 10.1109/access.2022.3198973.

Kshetri, Nir. "1 Blockchain's roles in meeting key supply chain management objectives." *International Journal of Information Management* 39 (2018): 80–89.

Kumar, Amit, Maria Holuszko, and Denise Crocce Romano Espinosa. "E-waste: An overview on generation, collection, legislation and recycling practices." *Resources, Conservation and Recycling* 122 (2017): 32–42.

Lata, Kusum, and Shri SK Singh. "IOT based smart waste management system using Wireless Sensor Network and Embedded Linux Board." *International Journal of Current Trends in Engineering & Research* 2, no. 7 (2016): 210–214.

Lim, Zhe Xi, Sasidharan Sreenivasan, Yew Hoong Wong, and Kuan Yew Cheong. "Aloe vera in active and passive regions of electronic devices towards a sustainable development." In *AIP Conference Proceedings*, vol. 1865, no. 1, p. 050006. AIP Publishing LLC, 2017.

Muniswamaiah, Manoj, Tilak Agerwala, and Charles C. Tappert. "Green computing for Internet of Things." In *2020 7th IEEE International Conference on Cyber Security and Cloud Computing (CSCloud)/2020 6th IEEE International Conference on Edge Computing and Scalable Cloud (EdgeCom)*, pp. 182–185. IEEE, 2020.

Okewu, Emmanuel, Sanjay Misra, Rytis Maskeliūnas, Robertas Damaševičius, and Luis Fernandez-Sanz. "Optimizing green computing awareness for environmental sustainability and economic security as a stochastic optimization problem." *Sustainability* 9, no. 10 (2017): 1857.

Pan, Xu, Wong, Christina W.Y., & Li, Chunsheng (2022). Circular economy practices in the waste electrical and electronic equipment (WEEE) industry: A systematic review and future research agendas. Journal of Cleaner Production, 365, 13267110.1016/j.jclepro.2022.132671.

Patil, Seema S., Smita J. Ghorpade, and Ratna S. Chaudhari. "Need of collaborative work towards emerging issues green computing and green IoT." (2018).

Patwa, Nitin, Uthayasankar Sivarajah, Arumugam Seetharaman, Sabyasachi Sarkar, Kausik Maiti, and Kunal Hingorani. "Towards a circular economy: An emerging economies context." *Journal of Business Research* 122 (2021): 725–735.

Pieroni, Alessandra, Scarpato, Noemi, Di Nunzio, Luca, Fallucchi, Francesca, & Raso, Mario (2018). Smarter City: Smart Energy Grid based on Blockchain Technology. International Journal on Advanced Science, Engineering and Information Technology, 8, 29810.18517/ijaseit.8.1.4954.

Rai, Rahul, Manoj Kumar Tiwari, Dmitry Ivanov, and Alexandre Dolgui. "Machine learning in manufacturing and industry 4.0 applications." *International Journal of Production Research* 59, no. 16 (2021): 4773–4778.

Ramadoss, Tamil Selvan, Hilaal Alam, and Ramakrishna Seeram. "Artificial intelligence and Internet of Things enabled circular economy." *The International Journal of Engineering and Science* 7, no. 9 (2018): 55–63.

Ramanujam, V., and D. Napoleon. "IoT based green computing: An attempt to delineate e-waste management." (2020).

Ramesh, Maneesha V., K. V. Nibi, Anupama Kurup, Renjith Mohan, A. Aiswarya, A. Arsha, and P. R. Sarang. "Water quality monitoring and waste management using IoT." In *2017 IEEE Global Humanitarian Technology Conference (GHTC)*, pp. 1–7. IEEE, 2017.

Rao, L. Nageswara. "Environmental impact of uncontrolled disposal of e-wastes." *International Journal of ChemTech Research* 6, no. 2 (2014): 1343–1353.
Recereum. 2018. Available from: http://recereum.com/.
Reuter, M., and A. Van Schaik. "Gold–A key enabler of a circular economy: Recycling of waste electric and electronic equipment." In *Gold Ore Processing*, pp. 937–958. Elsevier, 2016.
Roychoudhuri, Reshma, and Biswajit Debnath. "Risk assessment of e-waste recycling with a focus on ISO 9000 standard." In *Sustainable Waste Management: Policies and Case Studies*, pp. 431–440. Springer, Singapore, 2020.
Sahoo, Swagatika, and Raju Halder. "Blockchain-based forward and reverse supply chains for E-waste management." In *International Conference on Future Data and Security Engineering*, pp. 201–220. Springer, Cham, 2020.
Schafer, Burkhard. "D-waste: Data disposal as challenge for waste management in the Internet of Things." *The International Review of Information Ethics* 22 (2014): 101–107.
Schluep, Mathias. "Waste electrical and electronic equipment management." In *Handbook of Recycling*, pp. 397–403. Elsevier, 2014.
Sengupta, Gaurav. "Green computing–new perspective of efficient usage of energy and reduction of e-waste." *Journal of Technology and Innovation in Tertiary Education* 2, no. 1 (2019): 11–16.
Sensa Networks. "How is blockchain tech impacting waste management? from bitcoins to environmental protection." (2019). Available from: http://sensanetworks.com/blog/how-is-blockchain-tech-impacting-waste-management-from-bitcoins-to-environmental-protection/.
Surapaneni, Praneetha, L. Maguluri, and P. M. Syamala. "Solid waste management in smart cities using IoT." *International Journal of Pure and Applied Mathematics* 118, no. 7 (2018): 635–640.
Swachhcoin. (2017). Available from: https://swachhcoin.com/.
Tao, Chen, and Li Xiang. "Municipal solid waste recycle management information platform based on internet of things technology." In *2010 International Conference on Multimedia Information Networking and Security*, pp. 729–732. IEEE, 2010.
Ullah, M., and Biswajit Sarkar. "Smart and sustainable supply chain management: A proposal to use rfid to improve electronic waste management." In *Proceedings of the International Conference on Computers and Industrial Engineering, Auckland, New Zealand*, pp. 2–5. 2018.
Wang, Cong, Jiongming Qin, Cheng Qu, Xu Ran, Chuanjun Liu, and Bin Chen. "A smart municipal waste management system based on deep-learning and Internet of Things." *Waste Management* 135 (2021): 20–29.
Wang, Lihui, Xi Vincent Wang, Liang Gao, and József Váncza. "A cloud-based approach for WEEE remanufacturing." *CIRP Annals* 63, no. 1 (2014): 409–412.
Wang, Xi Vincent, and Lihui Wang. "WRCloud: A novel WEEE remanufacturing cloud system." *Procedia CIRP* 29 (2015): 786–791.
Wath, Sushant B., Vaidya, Atul N., Dutt, P.S., & Chakrabarti, Tapan (2010). A roadmap for development of sustainable E-waste management system in India. Science of The Total Environment, 409, 19–3210.1016/j.scitotenv.2010.09.030.
Widmer, Rolf, Oswald-Krapf, Heidi, Sinha-Khetriwal, Deepali, Schnellmann, Max, & Böni, Heinz (2005). Global perspectives on e-waste. Environmental Impact Assessment Review, 25, 436–45810.1016/j.eiar.2005.04.001.
Zhang, Kejing, Ping Cang, Jutta Geldermann, and Fugen Song. "Research on innovative information-flow management of e-waste recycling network based on cloud computing." In *2010 Chinese Control and Decision Conference*, pp. 1049–1053. IEEE, 2010.

3 Printed Circuit Board Recycling

Anthony Phiri,
Quinton Chamunorwa Kanhukamwe,
and Tichaona Goto

3.1 INTRODUCTION

Electronic waste (e-waste), e-scrap, or waste electrical and electronic equipment (WEEE) generally refers to discarded electrical or electronic devices. From a broader perspective, e-waste may also be defined as discarded computers, office electronic equipment, entertainment device electronics, mobile phones, television sets, and refrigerators, which are made of sophisticated blends of plastics, metals, and other materials (Nagajothi and Felixkhala 2015; Seebeger et al. 2016). The majority of office and household gadgets have electronic materials. Discarded electronic waste is the fastest-growing stream of waste in industrialized countries (Toxics links, 2004). The market for electronics might be reaching saturation; therefore, the third world is the next recipient. Nearly 54 million metric tons of e-waste were generated worldwide in 2019 (Statista 2019). The rapid global modernization resulting in rapid urbanization, population growth, and the improving global economy, has led to unprecedented levels of end-of-life electronic waste, or printed circuit board (PCB). The challenges in PCB waste management are disproportionately acute in third-world countries than in developed countries. The fundamental difference is caused by the absence or lack of sound policies, frameworks, acts, and implementation strategies in managing typical waste. Generally, third-world countries have inapplicable or no relevant policies and supporting infrastructure, leading to their struggle to launch a sustainable waste management system. Countries like Nigeria, which have a high influx of electronic gadgets, are failing to implement international environmental treaties due to lack of a national policy framework and programme to implement them. As a result, a majority of water bodies in Nigeria are polluted and not fit for human consumption. The main reason is the unregulated disposal of e-waste and consequently generate metals. Kenya is faced with a myriad of challenges that include low citizen awareness, lack of proper policy and legislative framework, disposal laws, and inadequate infrastructure among others. In Nepal, the informal sector is the sector that deals with electronic waste, but the lack of understanding of the existing system hinders any advancement in this sector. Although South Africa has good legislative frameworks that strongly recognize the principles of environmental protection and rehabilitation; however, there is no specific legislation dealing with e-waste.

DOI: 10.1201/9781003317050-4

In India, various issues emanated due to first economic liberalization, i.e., after 1990 (Manasi 2013). Due to the scrapping of import restriction laws in 1990, there was an increase in consumption of electronic waste.

A new growth in the service sector was noticed due to infrastructural reforms and the globalization process in the 1990s. In addition to that, the technological advancement resulted in economic electronic products that were user friendly as well. The process led to an exponential increase in the consumption pattern contributing to e-waste pilling (World Information Technology and Services Alliance (WISTA) 2002). According to Khurrum et al., 2011, the negative side is that the lifespan of these products became increasingly shorter and the accumulation of electronic waste brought another pollution problem. The obsolescence of these products leads to a unique mindset where consumers preferred to replace the products rather than repair and reuse (Nagajothi and Felixkhala 2015). The rapid obsolescence is also due to the rapidly evolving technology but, on the other hand, it is clear that the throw-away principle yields monetary benefits to corporate rather than the technology users. Small gadgets and large gadgets in China have PCBs, and often they are not reparable. This is a major driver of increase of e-waste in the communities.

Generally to extract value from waste streams, each waste stream has to be managed in a specific way, which ensures waste resources and energy conservation. From the picked case studies, it is clear that each country has specific challenges in managing electronic waste.

These are not limited to:

- Lack of environmentally conscious design and development trends
- Lack of waste-specific management approaches
- Lack of policy and legislative frameworks for e-waste management
- Inadequate waste management resources and infrastructure

The need to manage e-waste appropriately cannot be over emphasized, given the polluting potential. It is the largest growing waste stream in the world, and its output parallels economic development. E-waste is qualified to be described as hazardous waste, as it contains organic and inorganic compounds that pose a harmful impact on the environment. The understanding of the impact of poor waste management is critical. Some people cannot easily understand the environmental impact of a piece of electronic waste item or machinery when it is lying in the environment. Generally people have to be conscientized that the composition of an item makeup is what makes it hazardous. At the end of those items' lifespan, appropriate handling is critical, and it can either result in controlled environmental impact or none. The appropriate management will result in both human and environmental protection.

3.2 ELECTRONIC WASTE RECYCLING TECHNOLOGIES

Any electronic or electrical gadget contains an electronic circuit board, which is the PCB. The materials that make the PCB differ in type, quantity, and concentration, and this also dictates the appropriate method of recycling the gadget. PCBs are composed of metal, polymers, and glass fibre, and they are mass-produced by different technologies.

Due to the escalating production costs, the need to protect the environment, and the desire to ensure sustainable development, the focus on end-of-life management has increased. Initiatives by various countries to improve sustainability by promulgating appropriate policies with clear targets is one of the solutions. The disposal of PCBs in landfills has been considered as the last resort, under sustainable development approaches, as this throws away potential resources. Such approaches have been considered as waste of waste resources and environmental pollution. In essence, recycling of waste PCBs is critical as this enables recovery of valuable waste that can be kept in products' life cycle and reduces over dependence on virgin material. Due to variation in composition, resource recovery and recycling of PCBs requires an inter-disciplinary approach to take care of each type of material type and avoid unintended environmental pollution as recyclers tend to pick specific material for their specific needs, leaving other materials in the environment. Among other issues, this section unpacks some of the resource recovery approaches.

3.2.1 Recyclable Material

Almost all the various materials that compose the PCB are capable of being recycled. PCBs have variation in type, concentration, and quantity of material in each specific gadget. PCBs are a base foundation on which electronic components and connections are mounted (Luda 2011). Typical composition of PCBs is metals 40%; polymers 30%, and ceramics 30% (Luda 2011). A range of metals are contained in PCBs. These are not limited to metals such as aluminium, zinc, copper, gold, palladium, nickel, and iron. The high-value metals contained are gold and palladium, which are often in very low quantity and concentration. The board laminate predominantly consists of a glass fibre reinforced thermosetting matrix. For easier understanding of PCB recycling, they shall be grouped into two groups: metals and non-metals. The non-metal component is a polymer matrix with reinforcement. The bare board is made of glass fibre reinforced thermosetting composite compounds. Unlike thermoplasts, thermosets can't be melted down, so they usually have to be broken down, mixed with an adhesive, and then pressed back together. The woven glass fibre is often used as the adhesive. Although any other material, especially light and strong, can be used in PCB reinforcement, predominantly glass fibre is used. Often the starting point in processing recyclables in PCBs is crushing. In crushing, broken fibres are recovered as short strand fibres. The recovered fraction retains a high length-to-density ratio, good elastic modulus, and low elongation, which ensures they can be reused. This means there is high potential of use of recovered material.

The other group is the metals group. The PCBs have both precious and non-precious metals in varying quantities. A PCB contains metals in different amount and ratios. A standard PCB contains iron 5%, copper 27%, aluminium 2%, and nickel 0.5%, whereas gold and silver are present in trace levels, 2,000 ppm and 80 ppm, respectively (Luda 2011). Gadgets of the same type that perform common or the same service in a PCB may vary in matrix composition due to brand, manufacturer, or year of manufacturing. This explains why there is no average scrap composition. This means the above given average values of composition are typical averages and only represent scraps of a certain age and manufacturer. Current trends show that the metal content in PCBs has gradually gone down due to the falling power consumption of modern switching circuits. Historically, in the 1980s the contact layer was 1–2.5 μm thick; in modern appliances, it is between 300 and 600 nm. This reflects the future possible value reduction of recyclables in the PCB recycling business as technologies advances.

3.2.2 Recycling Technologies

The recycling process is appropriately done in a systematic way starting with the sorting process. The sorting stage enables the recovery of parts of the PCB that can be used without any physical or chemical change. This effectively preserves the value of the particular component and conserves energy. Valuable components like cables and engineering plastics can be recovered in reusable condition. Items such as cell batteries and capacitors can be manually removed for reuse or separately disposed in an appropriate way. After the sorting stage, the PCBs can then be sent to a facility for further dismantling for reuse or reclamation of electric components. To salvage items in reusable condition, manual dismantling is appropriate. The latest trends show that the use of image processing and database to recognize reusable parts or toxic components can soon be the way forward. In a way, this technology will enable precise selection and recovery of reusable parts of PCB with high precision. The disassembly by use of image processing and database to recognize reusable parts or toxic components is very appealing, especially where there is a large e-waste quantity. The automated disassembly of electronic equipment is well advanced, but unfortunately, its application in recycling of electronic equipment still faces a lot of frustration.

Reusable electronic components have to be dismantled from PCB assembly as the most important step in their recycling chain, to ensure conservation of energy and material resources, reuse of components, and elimination of hazardous materials from the environment. In semi-automatic approaches, electronic components are removed by a combination of heating and application of impact, shearing, and vibration forces to open soldered connections. Heating temperature of 40–50°C higher than the melting point of the solder is necessary for effective dismantling (Duan et al. 2011).

3.2.2.1 Physical Recycling

In physical recycling, the first stage is size reduction followed by separation of metallic and non-metallic fractions and followed by further management for specific use. In this approach, the small fractions consisting of thermosetting resins, glass fibres or cellulose paper, ceramics, and residual metals can be used as filler material for different resin matrix composites. Shredders, cutting mills, and granulators are used for size reduction, and sieves are used at an elevated temperature. Temperatures can be sustained to around 250°C so as to enable pyrolytic cleavage of chemical bonds in the matrix. The majority of mechanical recycling processes have a certain effective size range, and mechanical separation processes are performed using a variety of techniques to ensure the particles can be used for a specific purpose. Shape separation by tilted plate and sieves is the most basic method that has been used in the recycling industry. Magnetic separators, low-intensity drum separators, are widely used for the recovery of ferromagnetic metals from non-ferrous metals and other non-magnetic wastes. The use of high-intensity separators makes it possible to separate copper alloys from the waste matrix. Electric conductivity-based separation such as Eddy current separation, corona electrostatic separation, and triboelectric separation separates materials of different electric conductivity such non-ferrous metals from inert materials (Cui and Forssberg 2003; Veit et al. 2005). Density-based separation of particles such as sink-float separation, jigging, and upstream separation is also used to separate metal from non-metal fractions in PCB scraps.

The non-metal PCBs parts have high thermal stability and are suitable for the moulding process. The thermosetting matrix is effective for making composites with PCB scraps and as reinforcement fillers. Predominantly, the 300–700°C pyrolysis residues (75–80%) can be used as filler materials as well as to replace wood flour in production of wood plastic. Non-metallic waste can also be used in concrete structures and asphalt construction. It is effective as a filler substitute and effectively decreases the dead weight of structures.

3.2.2.2 Chemical Recycling

Chemical recycling is the decomposition of the waste polymers into their monomers or some useful chemicals by means of chemical reactions (Luda 2011). Some of the notable chemical recycling approaches are pyrolysis process, depolymerization, hydrogenolytic degradation, and gasification process.

3.2.2.2.1 Pyrolysis and Dehalogenation

In pyrolysis, a polymer breaks down in the absence of oxygen to produce an oil/wax, a gas, and a char product, leaving a solid residue. Pyrolysis has the advantage that potentially all of the products from the process can be used. In the process, glass fibre is recovered from composite plastic waste; oil/wax gas is produced and can be used as a liquid fuel or as a chemical feedstock to produce new plastics. It is a cost-effective recycling process, effective sustainable manufacturing. This could be an answer to the European End of Life Vehicle Directive as the thermosets will be usefully converted. The pyrolysis process gives an opportunity for wide use of thermosets, especially in lightweight components. E-waste plastics contain toxic halogenated flame retardants, which may cause serious environmental pollution. This is prevalent during treatment in pyrolysis as they form carcinogenic substances polybrominated dibenzo dioxins/furans (PBDD/Fs). The effect can be reduced by dehalogenation. Dehalogenation and pyrolysis of plastics can be carried out simultaneously or successively. The first strategy essentially is the two-stage pyrolysis with the release of halogen hydrides at low pyrolysis temperature region, which is separate from the decomposition of polymer matrixes, thus obtaining halogenated-free oil products

(Yang et al. 2013). The second strategy is the most common method. Zeolite or another type of catalyst can be used in the pyrolysis process for removing organohalogens (Yang et al. 2013). The third strategy separates pyrolysis and dehalogenation of WEEE plastics, which can, to some degree, avoid the problem of oil value decline due to the use of a catalyst, but obviously, this strategy may increase the cost of the whole recycling process (Yang et al. 2013).

3.2.2.2.2 Gasification

In the gasification process, plastic waste is reacted with a gasifying agent (e.g., steam, oxygen, and air) at a high temperature of around 500–1300 °C, which can produce synthesis gas or syngas. Syngas is an intermediate product composed of carbon monoxide (CO) and hydrogen (H_2) and, by adding steam, can be converted to carbon dioxide and hydrogen. By adding steam and reacting over a catalyst in a water-gas-shift reactor, the mixture burns, producing heat and water. This combustion can and is being used for creating electricity without any greenhouse gas generation. The current drive to use hydrogen widely for power generation creates big business in recycling PCBs by the gasification process as a clean energy source. Hydrogen-enriched syngas can be used to make gasoline and diesel fuel, and captured carbon dioxide from syngas can be stored and used in various processes where it's required.

3.2.2.2.3 Biodegradable Plastics

Plastics are a menace to the environment, and the best way to reduce accumulation, improve resource efficiency, and reduce the environmental impact of plastics is the prevention of waste. Substitution of non-biodegradable plastic by degradable plastics is the most effective way. Biodegradable plastics are capable of contributing to a sustainable society and contribute to the reduction in carbon dioxide emissions during production. Biodegradable plastics offer new end-of-life management options that have a lower or no negative impact on the environment and create opportunities for increased use of plastic material, which is comparatively strong but cheaper than other construction material. Biodegradable polymers include polylactic acid (PLA), thermoplastic starch (TPS), polyhydroxyalkanoate (PHA), polycaprolactone (PCL), and poly (butylene adipate-co-terephthalate) (PBAT). Biodegradable plastics are not toxic, which means extracting metals from PCB becomes easier and cheaper when biodegradable plastics can be naturally done away with.

3.3 STUDY METHODOLOGY

3.3.1 METHODS

One specific methodology was used to gather the various information sources and data. During the data acquisition period, the investigators developed a selective perception of the entire e-waste recycling sector. The method practically took an approach of gathering information through online literature survey, government-published data, and various companies' catalogues. The above approach is applied to various countries and regions to bring out comparisons and specifically on the way they handle e-waste. To discern the trends through an overview of important information and data that should be obtained during the data acquisition phase, a summary of sources of information was developed in a descriptive manner and tabular.

3.3.2 Materials

Various materials were used in this literature survey. The material included but was not limited to books, manuals, periodicals, scholarly articles, and any other sources relevant to e-waste and, by so doing, provided a description, summary, and critical evaluation of these works in relation to the e-waste issue.

From a broader perspective, the sections can be broken down into the following categories:

- Literature survey: internet, database, specific reports, and press
- Statistical data: national statistics, census, and databases
- Companies' catalogues

The predominant e-waste management approaches were identified, and the gradual gravitation towards certain approaches was noted and presented in the results section. Deductions on current trends were made based on the collected information.

3.4 RESULTS AND DISCUSSION

A basic PCB consists of a flat sheet of insulating material and a layer of copper foil, laminated to the substrate. Chemical etching divides the copper into separate conducting lines called tracks or circuit traces, pads for connections, vias to pass connections between layers of copper, and features such as solid conductive areas for electromagnetic shielding or other purposes (Sparkfun 2020). The tracks function as wires fixed in place and are insulated from each other by air and the board substrate material. The surface of a PCB may have a coating that protects the copper from corrosion and reduces the chances of solder shorts between traces or undesired electrical contact with stray bare wires. For its function in helping to prevent solder shorts, the coating is called solder resist or solder mask. From the observed design of the PCB, it's clear that there is an entanglement of bits and pieces, and extricating any of them is not easy.

According to Sparkfun (2020), electronics became more prevalent in consumer goods, and the pressure to reduce the size and manufacturing costs of electronic products drove manufacturers to look for better solutions. That resulted in the PCB used today. Besides the great technical advantage that was developed, the challenge was left in disposal approaches given the big volumes. The issue of size reduction was primarily focussed on reducing the expenses associated with manufacturing. The manufacturing costs were not viewed from the recycling perspective, and if the manufacturers had added the recycling costs, size wouldn't have been an issue but rather functionality.

According to Gui and Forssberg (2003), WEEE is diverse and complex in terms of materials and components as well as the manufacturing process. Characterization of this waste stream is critical for developing a cost effective and environmentally friendly recycling system. While modern PCBs are versatile, there's always room for development. In order to deliver new features that connect both consumers and professionals, product manufacturers are focusing on extracting even more power from printed circuit boards. It is clear that with advancement of technology comes complexity in producing and disposing of it. Due to its complex composition, PCB recycling requires a multidisciplinary approach intended to valorize fibres, metals, and plastic fractions and reduce environmental pollution, which are here reviewed in an attempt to offer an overview of the latest results on recycling waste PCBs. According to Gui and Forssberg (2003), typically PCBs contain 40% metals, 30% organics, and 30% ceramics. However, there is a great variance in composition of PCB wastes coming from different appliances, from different manufacturers, and from different-age gadgets. For recyclers, it could be important to profile gadgets, for example, according to age so that recycling activities are focused on particular items in abundance with understanding of the potential value in the particular e-waste type. As an example, after removing hazardous batteries and capacitors which, according to current legislation, must follow separate recycling, the organic fraction resulted in about 70% in PCBs from computers and TV sets and 20% in those from mobile phones (William and Williams 2007).

The issues of PCB milling and PCB made from biodegradable material can be the founding principles for a policy that supports a circular economy approach. While milling is the process of removing areas of copper from a sheet of printed circuit board material to recreate the pads, signal traces, and structures according to patterns from a digital circuit board plan, PCB from biodegradable material is self-explanatory as unpacked in the conclusion. It would seem appropriate to have a policy that promotes a hierarchy of e-waste management whereby emphasis is on the use of biodegradable material, followed by a milling approach, and lastly by general component salvaging for reuse.

Generally, the management of e-waste is done by the formal and informal sectors, and they handle waste differently and for different purposes. Not all parts of electronic gadgets become useless after one use. Some remain reusable in that original state. So there is a tendency by various organizations to salvage the easily reusable parts for use in other gadgets. This can be done in formal or informal sectors. The remaining material can now be discarded, and the informal sector also starts to salvage materials as well. It is clear that there is a sustainable value chain of activities. As an intervention to make e-waste management sustainable, a value chain can be mapped based on material flow from the formal sector to the informal sector. The value chain analysis can be bottom-up or top-down. That is to say, electronics manufacturers release gadgets to distributors and then distribute them to salvagers of specific items and then to the informal sector, who salvage little in terms of value. Or vice versa.

Having dwelt on the various aspects of waste, the following part dwells on recommendations for some of the available options in managing e-waste that the literature survey has reflected. Basically, the enabling environment is divided into two parts and within them are sub-activities or approaches. The chapter recommends that salvaging (reuse/recycle, incineration) value and a legislative environment can complement each other very well to extract maximum value from e-waste.

According to Nagajothi and Felixkhala (2015), a policy is appropriate to address all issues ranging from production to final disposal, including environmentally sound technology for the recycling of electronic waste. They further indicate that regulations should be able to control both legal and illegal exports and imports of e-wastes and must be clear in the policy. Extended producer responsibility is critical as well as banning electronic waste at landfills. Such an approach ensures all electronic waste salvaging activities are implemented within the legal and policy framework. Lastly formalizing the informal sector brings high benefit to disadvantaged communities (Manasi 2013). E-waste processing can be a source of livelihood for many poor families. They can be leveraged by formally training them in e-waste processing and authorizing them to do so by the policy.

3.5 CONCLUSION

Among other constituents, e-waste contains many components such as mercury, lead, cadmium, polychlorinated bi-phenyls, brominated flame retardants, and etched chemicals. These chemical constituents exist in varying quantities in each gadget and also in the age of technology. Technology advancement within the field of PCB is happening at a tremendous speed. This makes it hard to keep up with the best ways in e-waste management technologies. Worse still, assessment of the development to acquire evidence is becoming more difficult. This is why the literature review as a research method was used and is more relevant than ever. The chapter defines and analyses the electronic waste management advances, while offering a broader analysis of the relevant literature in order to summarize the information available and determine a technology development pathway. Based on this, a few key points were realized. Firstly, many countries don't have any standardized method to estimate e-waste generation and management. This suggests that there is a need to develop and implement policy frameworks for appropriate e-waste management in developing countries so as to solve environmental issues related to informal recycling practice. There is a need for developing a legal framework for the management of this waste fraction that supports economic and sustainable value chain management improvement, especially in developing countries. The policies and legal framework should enable development of programmes that ensure training is provided in handling e-waste.

The world over, due to dwindling material, researchers and manufacturers have focused on producing gadgets that have little material uptake, have high processing capacity, and are biodegradable. This has resulted in slimming down of electronic gadgets being predominant. Durability and recyclability still remain aspects that are critical. As a result of efficient manufacturing processes and innovative design software, modern technology focuses on smaller, lighter, and more powerful devices.

While the technology to develop slimmer designs keeps on advancing, the environmental concern has received less attention, but international standards demand compliance. Of late biodegradable printed circuit boards have been developed using bio-composites (Crimp Circuits 2020). These bio-composites are chemical free and are normally considered as agricultural wastes or co-products; therefore, they can be co-disposed at a landfill with other biodegradable waste. Natural fibres and protein-based PCBs provide an excellent alternative to the synthetic polymer-based circuit boards. It helps us to reduce the environmental burden due to the disposal of electronic waste (e-waste) (Crimp Circuits 2020). The use of biodegradable PCB can enable advances in technology while safeguarding the environment.

REFERENCES

Crimp Circuits. 2020. Advancement and future of printed circuit board. https://www.crimpcircuits.com/blog/advancement-future-of-printed-circuit-board.

Cui, J & Forssberg, E. 2003. Mechanical recycling of waste electric and electronic equipment: a review. *Journal of Hazardous Materials*, 99(3): 243–263.

Duan, H., Hou, K., Li, J., & Zhu, X. (2011). Examining the technology acceptance for dismantling of waste printed circuit boards in light of recycling and environmental concerns. *Journal of Environmental Management*, 92(3): 392–399.

Gui, J. and Forssberg, E. 2003. Mechanical recycling of waste electric and electronic equipment. *Journal of Hazardous Materials*, 99: 243–263. 10. 1016/s0304-3894(03)00061-x

Khurrum, M.S., Bhutta, I, Adnan Omar, A and Yang, X. 2011. Electronic waste: a growing concern in today's environment. *Economic Research International*. https://www.hindawi.com/journals/ecri/2011/474230/

Luda, M.P. 2011. Recycling of printed circuit boards. *CINECA IRIS Institutional Research Information System*. http://www.intechopen.com/articles/show/title/recycling-of-printed-circuit-boards http://www.intechweb.org

Manasi, S. 2013. Emerging trends in e-waste management status and issues a case study of Bangalore city. *Institute for Social and Economic Change*, 10.13140/RG.2.2.27490.32964

Nagajothi, P. G. and Felixkhala, K. 2015. Electronic waste management: a review. *International Journal of Applied Engineering Research*, 10(68). https://www.ripublication.comijaer.htm.

Seebeger, J., Grandhi, R., Kim, S.S., Mase, W.A., Reponen, T., Mei-Ho, S. and Chien, O. (2016). Special report: e-waste management in the United States and public health implications. *Journal of Environmental Health*, 79 (3): 8–16.

Sparkfun. 2020. Printed circuit boards (PCB) basics. https://learn.sparkfun.com/tutorials/pcb-basics/all. Accessed 07/08/2020.

Statista. 2019. Electronic waste generated worldwide from 2010 to 2019 (in million metric tons). *Journal of Energy & Environment* https://www.statista.com/statistics/499891/projection-ewaste-generation-worldwide/

Veit, C et al. 2005. Ageing mechanisms in lithium-ion batteries. *Journal of Power Sources*, 147(1–2), 9 September 2005; Pages 269–281.

William, J. H. and Williams, P. T. 2007. Separation and recovery of materials from scrap printed boards. *Resources, Conservation and Recycling*, 51: 691–709, 10.1016/j.resconrec.2006.11.010 ISSN: 09213449.

World Information Technology and Services Alliance (WITSA). 2002. Digital Planet 2002 Global Information Economy, Vienna, VA 22182 USA.

Yang, X et al. 2013. Pyrolysis and dehalogenation of plastics from waste electrical and electronic equipment (WEEE): a review. *Waste Management*, 33(2), February 2013: 462–473.

Part 2

Smart and Intelligent Systems

4 Decision Tree Approach for Classifying E-waste to Recycle and Reproduce

Anubha Dubey and Apurva Saxena Verma

4.1 INTRODUCTION

In a very simple way, electronic waste (e-waste) is those electronic products that are unwanted and not in use like computers, televisions, VCRs, stereos, copiers, fax machines, and mobiles. Office and medical instruments also come under this category. Persons throw these products out if broken or not in use; donate them; or sell them. If electronic items are not selling in shops, people throw them away. E-waste is menacing due to hazardous chemicals that naturally leach from the metals inside when buried. Simply, it is called 'digital waste,' or e-waste. These modern amenities contain some form of poisonous materials, including beryllium (Be), cadmium (Cd), mercury (Hg), and lead (Pb), which cause environmental threats to the soil, water, air, and wild flora and fauna. And when they end up in landfills, their microscopic traces dissolve in sludge and permeate the landfills. As more and more e-waste and metals are ending up in landfills, more toxic materials are showing up in groundwater. Improper disposal of e-waste leads to many life-threatening circumstances for humans and animals; ultimately our environment gets effected (Devika 2010). There are so many e-waste materials that can be categorized (Miliute-Plepiene and Yuhana 2019) as:

(a) Equipment that can change temperature: our daily use appliances like refrigerators, freezers, and heat pumps.
(b) Lamps/electric items: street fluorescent lamps, tube lights, table lamps, and LED lights.
(c) Large tools: washing machines, clothes dryers, cookers, musical equipments, sports equipment, and medical devices.
(d) Telecommunication appliances: mobile phones, global positioning systems (GPSs) and navigation tools, pocket calculators, printers, and telephones.
(e) Small daily equipment: vacuum cleaner, microwaves, and thermostat.

These appliances are made up of metals, rare earth metals, elements, plastics and petroleum-based substances, metallic and non-metallic materials, and hazardous substances. Gold, silver, platinum, palladium, copper, nickel, tantalum, cobalt,

DOI: 10.1201/9781003317050-6

aluminium, tin, and zinc are some names of metals and rare earth elements used in different electronic devices (Bakas et al. 2016, Chen et al. 2011). In this technology era, demand for these devices is increasing day by day, which is exceeding the limits of supply, and several substances are marked as critical; hence, if demand increases like this, these critical elements will no longer be available. These elements are extracted from all over the world, and due to their efficient properties, are used in designing and manufacturing of electrical and electronic tools. The extraction methods for these elements are also sometimes low quality and hazardous. It is recommended that all countries properly follow safety precautions for the safety of workers in mines for these elements.

All these products have their life cycle from material extraction to waste treatment. For example, a smart phone (Baldé et al. 2017) consists of different metals, plastics, ceramics, and other trace materials. These stages also have air ejections, outflow, and debris. The production of microchips, screens, and batteries is also resource-intensive and generates lots of waste. This needs transport demands as materials are coming from different parts of the world. Now, energy for electrical gadgets charging and utilizing Internet infrastructure is also required for facts and figures as well as data storage in the cloud. Understanding the life cycle of all the electronic gadgets is important while recycling, dismantling, or treating the single components. The value chain requires a systemic analysis of the life cycle of products.

4.1.1 Our Contribution Towards the Environment

We should become environment friends. All the rules and regulations need to be followed by all the companies and hospitals. It is very important to protect our groundwater from these toxic metals.

Solutions:

Recycling electronics: With the advancement of technology, there are many methods of recycling. These can be donating, creating e-waste art, or recycling.

Donating electronics: If the electronic item is in good condition and we want to get rid of it, a good idea is to donate it. There are organizations like HP, Dell, Best Buy, and Goodwill that accepted these donated devices. If anyone decides to donate, some points need to be considered: do not upgrade new hardware or software; remove all the data from the device. This will make you an environment friend and prevent data breaching.

E-waste art: Being a little creative, we can make art that is fun and also raises awareness against e-waste. There are some artists like Gabriel Dishaw and Peter McFarlane who make a living creating e-waste art by using circuit boards, wiring, and various metal components.

Recycling: This is a good solution if recycling is possible; one very good example is the Environmental Protection Agency (EPA), which reprocesses one million laptops. The energy rescued matches the electricity consumed by more than 3,500 U.S. families in a year. Through recycling, 35 thousand pounds of copper, 772 pounds of silver, 75 pounds of gold, and 33 pounds of palladium can be recovered and used to create new cell phones.

If customers understand the proper way of recycling their electronic gadgets, then they never throw them in the trash (Fangchao Xu 2014).

Technology researchers are developing new devices every day, and we update our devices to show off or because of our mood. These technologies fulfil modern needs like in SMART homes. The smart city concept was developed to make technology a servant because of our changing lifestyles. As we discussed above, the materials in e-waste that need to be recycled are steel, glass, and plastics. The extractions of raw materials from these are also important.

4.1.2 Extraction of Substances

The hydrometallurgical, electrometallurgical, and bio-metallurgical processes are the processes that can be used to separate metal fractions from e-waste during preprocessing. If all the companies for e-waste work with bio-metallurgical, then there is less chance to affect the environment.

4.1.3 E-waste Transformation

There are three main steps for recycling of e-waste: (a) collection, (b) preproceesing, and (c) end processing (Baldé et al. 2017).

Collection: All the rules and policies must be followed by e-waste collection companies. There must be effective awareness among the public that recycling is a good solution. It is very important that all people know the importance of recycling and reuse of gadgets so they will not throw these items away. Collection of all electronic waste in a particular place is essential. These electronic constituents are sorted at a collection reserve, where usable components are taken back to the consumer supply chain.

Pre-treatment: In this stage, expired equipment is dismantled manually, and individual elements are tested and isolated. This process is only done by the collection company. In the early phase, wiring boards, drives, and other constituents are taken out. Hammer mills are used for shredding e-waste scrap in the mechanical process (Baldé et al. 2017). Mineral dressing is the process that separates metals and non-metals, using methods like screening, magnetic separation, eddy current (e.g., separating aluminium from glass), and density-based separation techniques (separation of materials based on mass density).

End processing: In this step, metals and non-metals are further processed. Many persons have studied recycling and utilization of non-metallic materials. Printed circuit boards (PCBs) contain more than 70% non-metallic substances like resins and glass fibres/fibre optics. Thermosetting resins cannot be remoulded due to their chain arrangement in shape (Lucier and Gareau 2019). Here, a flowchart (Figure 4.1) is shown to combine these essential methods to follow the recycling and reuse of materials and their components.

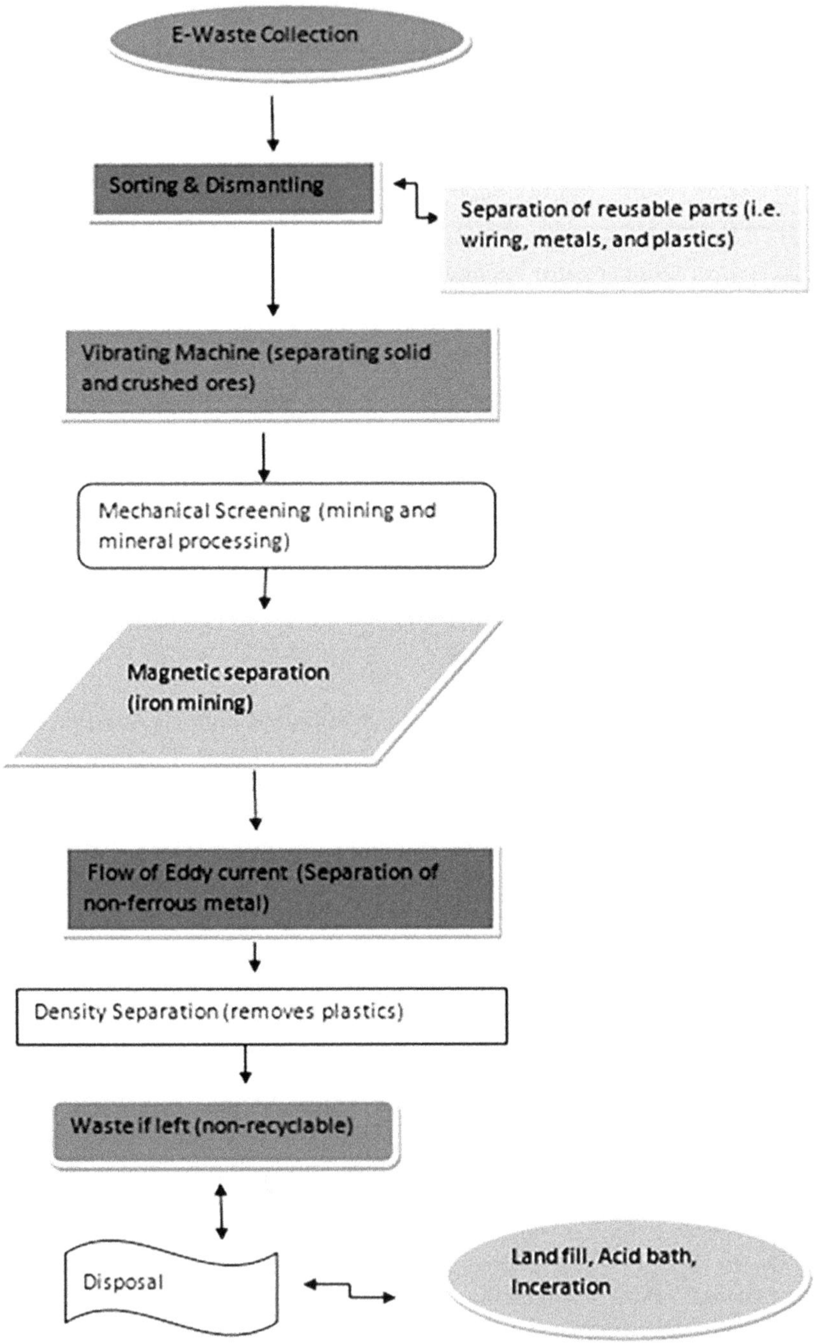

FIGURE 4.1 Flowchart of traditional e-waste separation and disposal.

4.2 RELATED WORK

Catania and Ventura (2014) explained a smart waste collection method. They tried to recover and optimize managing solid town waste material. The framework of a smart waste organization requires interconnection between heterogeneous procedures and data allotment relating to a large quantity of people. The Smart-M3 proposal resolves these problems, presenting a high level of decoupling and scalability. Gaidajis et al. (2010) tried to explain that environmental problems are increasing every day due to e-waste. Furthermore, the present and upcoming production of e-waste, along with the environmental troubles connected with its removal and management of performance, are explained in detail. According to Garcia-Garcia et al. (2019), manufacturing of food undergoes difficult processes that produce huge amounts of waste. They proposed a Waste Flow Model to accomplish two endeavours: (a) to utilize precious food from the manufacturer and waste data in an appropriate manner and (d) to examine present waste management of food to implement another solution of food salvage. Nguyen et al. (2018) explained the theory of planned behaviour (TPB), the structural equation modelling (SEM) used to study the important features that influence e-waste reuse behaviour. Data get evaluated on the following parameters: recycling, social pressure, laws and regulations, the cost of recycling, and the inconvenience of recycling, which significantly affected residents' behavioural intention, with laws and regulations. Esmaeilian et al. (2018) discussed the probability of stylish metropolitan and connected communities using waste management. Limitations of present waste management were defined, and a proposed structure of a waste management system was explained, along with the assessment of life cycle data in decreasing waste. Improvement in waste was enhanced, allowing the waste management system to connect with the whole life cycle product.

The aim of this chapter is to understand the various components by which electronic or electrical appliances are made, and then by the above-mentioned extraction method, these can be dismantled or reused for different purposes. The motivation behind writing this chapter is to introduce a decision tree for e-waste separation. This decision tree-based model is used for the classification of components used in any electronic device. If these can be recycled or reused, this helps in reducing human health hazards and is environmentally friendly.

4.3 MATERIALS AND METHODS

4.3.1 Sorting and Classification of E-waste

Trash collection is one form of waste management. It just shows that this trash will not affect the life of people and animals. But if it fills up landfills and enters oceans, the life of marine animals is in danger. In order to efficiently dispose of e-waste, sorting into biodegradable, recyclable, and non-biodegradable materials is important. Through sorting, it is possible to classify reutilization of biodegradable stuff to enrich the soil and remove harmful products. The recycling of waste substances like metal as well as paper and plastic is in high demand.

Since manual sorting is difficult and unhygienic, a smart system is maintained that classifies materials based on their characteristics through the machine learning decision tree concept.

A range of classification algorithms is used. For example, decision trees, Naïve Bayes probabilistic models, and neural networks are used to determine the best method for the development of the classification model (Dubey and Chouhan 2012). Here, we have tried to take three parameters to reduce electronic waste. They are metals, plastics, and glass because waste is categorized into these factors so that e-waste can be managed and utilized in the future. These methods are described as follows:

- *Metal:* This type of material is present in various items like televisions, refrigerators, cell phones, microwaves, washing machines, laptops, and tablets.
- *Plastic:* Material that is present in every electronic item.
- *Glass:* It is used in many items like automobiles and household goods.

All the data of the electronic waste material can help to categorize it into the three forms of metal, glass, and plastic, which are stored in the system. Now, our next move is to decide whether the waste material can be reused, recycled, or melted down. In this regard, the decision method of identification of the useful material from the waste is improved. We are in the age of cloud computing so our whole system is on the cloud to protect the data in the computing environment.

4.3.2 Dataset

Metal, plastic, and glass are the attributes that are obtained from the e-waste and further checked whether the item is useful and not useful according to its nature. With the help of the following table, we have made different combinations of test and train data simultaneously to generate the decision tree.

4.3.3 Classification of Substances

Recycling of substances like steel, glass, and plastic helps to reduce the dependency and extraction of raw materials from landfills. The new technology of machine learning is required to sort these substances without any manual tasks. Machine learning is a process where machines improve automatically through experience. It is a subset of artificial intelligence that takes a mathematical model in the form of an algorithm to implement. Of the various uses of machine learning, here we are learning about how to classify trash materials. Machine learning classification is a process of predictive modelling where a class label is predicted for a given example as input data. In this scenario, the algorithm works into four classes:

- Classify if the substance is reproducible.
- Classify if the substance is recyclable.
- Classify if the substance is reusable.
- Classify if the substance is melted down.

In classification, training datasets consist of inputs and outputs by which the machine learns how to use the data in a proper manner efficiently. The class labels are strings of values, e.g., substance *is reproducible or not reproducible,* which is called label encoding, where a unique integer is assigned to each class label, e.g., 'Reproducible'=0, 'non reproducible'=1.

4.3.4 Decision Tree

It is a supervised machine learning method based on the principle that the data are continuously split according to a certain parameter. It can be understood as:

Parts of the decision tree

Nodes: They test for a certain parameter (attribute).

Edges/branch: They correlate with the outcome of a test and connect to the next/corresponding node, which is called a leaf.

Leaf nodes: Nodes are terminal, that is, able to predict the outcome (showing class distribution).

As the decision tree algorithm is supervised in machine learning, the model is trained for the attributes used to find out the useful material. Then, it is required to calculate Entropy by class, Entropy by attribute, and information gain for designing the decision tree.

4.3.4.1 Designing of the Algorithm for the Decision Tree

- Step 1: With the existing dataset according to the requirements, the attributes (like metal, plastic, and glass) are decided. The last attribute of the dataset will become the class.
- Step 2: The entropy of the class attribute is calculated.
- Step 3: Calculate the entropy of the entire attribute, so that gain can be found.
- Step 4: Calculate the information gain of all the attributes of the dataset.
- Step 5: Calculate the gains of all the attributes. The maximum gain of the attribute will be the root node of the decision tree.
- Step 6: These steps are followed sequentially until the final decision tree is designed.

Entropy as it is understood is randomness and unpredictability in a number of monitoring systems. Entropy controls how a decision tree determines to split the data. It actually works on how a decision tree draws its frontiers, and information gain is used by decision tree algorithms to build an actual decision tree. Decision tree algorithms will always seek to maximize information gain. An attribute with the highest information gain will be tested/split first.

4.3.4.2 Evaluation of Decision Tree Generation

The entropy and information gain are the pillars of constructing a decision tree (Han and Kamber 2013).

Entropy (class): It is also called Shanon Entropy and is written as H(s) for a finite set *S*. It is an evaluation of the sum of unpredictability or randomness in data. Here *P* is the possibility of yes, and *N* is the possibility of a number in the dataset.

$$H(s) = -\frac{P}{P+N}\log_2\left(\frac{P}{P+N}\right) - \left(\frac{N}{N+P}\right)\log_2\left(\frac{N}{N+P}\right) \tag{4.1}$$

Information Gain: It is known as Kullback–Leibler divergence denoted by IG (S, A) for set *S*. It is the efficient change in entropy after a particular attribute, *A*. It calculates the relative change in entropy with reference to the independent variables.

$$I_{(P_i,N_i)} = -\frac{P}{P+N}\log_2\left(\frac{P}{P+N}\right) - \left(\frac{N}{N+P}\right)\log_2\left(\frac{N}{N+P}\right) \tag{4.2}$$

$$\textbf{Entropy (attribute)}\ E\left\{\frac{P_i + N_i}{P+N}\right\} \times I_{(P_i,N_i)} \tag{4.3}$$

$$GAIN = Entropy_{(class)} - Entropy_{(attribute)} \tag{4.4}$$

These calculations are able to find the attributes for making the final decision tree.

Hence in Figure 4.2, the decision tree suggested that out of all the parameters, the parameters metal and glass are the best parameters for the identification of reusable material or not. This is achieved by manual calculations also. Plastics are also part of reusable if they are durable, good quality, and used for the long term.

This decision tree is implemented in the collection and segregation of e-waste through smart dustbin. And it is very important to store the materials' data safe for further use and analysis. The decision tree is a well-established model for the categorization of materials based on their nature. It is the responsibility of all of us to protect nature and the environment from these harmful elements because ultimately, our life is dependent on nature directly or indirectly. The concept of smart dustbin is highly used as it will have the power of machine learning to collect all the biodegradable and non-biodegradable substances.

According to our model in Figure 4.3, the collector collects the e-waste and further divides it into metal, glass, and plastic. These materials are then stored in the smart dustbin because they are categorized through a smarter way like machine learning. After a smart dustbin, they get stored in the cloud provider services. This way we can provide security to the data and the algorithm on which categorization of the waste is possible. The rest of the waste material which comes under non-biodegradable is collected as waste and again goes in the processing stage. The data available to the cloud are used for storing and also for security. Because some of the hard metals that may harm the environment or any material that will harm nature are stored, without operator permission, no one can misuse them. Figure 4.3 is an eco-friendly model. Hence, as per the industry's protocol, these materials are sent to factories or industries for proper treatment

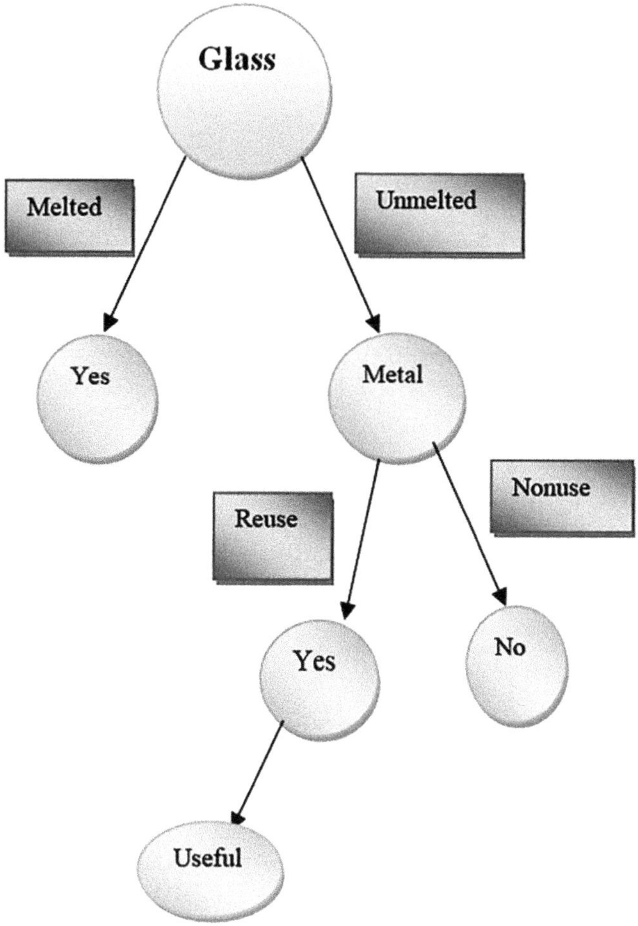

FIGURE 4.2 Final decision tree.

and reusable products. The landfill stage is where these wastes are buried under land following suitable rules and regulations. This process gives no harmful effect to the environment. On the other hand, all the hard materials like some metals iron, copper, etc. were processed as per their protocol, but bioleaching is also used. Bioleaching is an encouraging technique that makes use of the microbial world to recover metals from low-grade ores and e-waste (Hong and Valix 2014). This process has no harmful effects on the environment.

4.4 RESULTS AND DISCUSSIONS

The decision tree is found for the selected three attributes, metal, plastic, and glass. The entropy gains are calculated to check the performance of the decision tree. The smart-intelligent system for classification of e-waste is categorized, and the in-built calculation is fast according to set parameters.

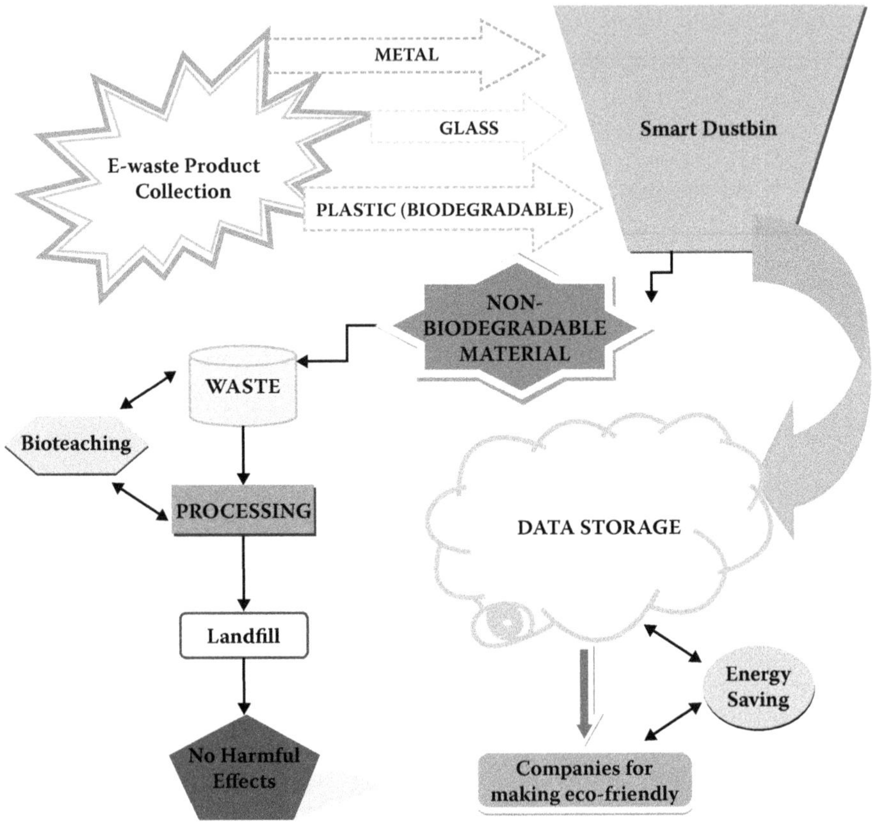

FIGURE 4.3 E-waste material categorization through smart and intelligent system.

The dataset in the form of Table 4.1 is used, and manual calculations are tried to form a decision tree to track the usefulness of waste. One of the attributes of Table 4.1, useful, is the class for entropy. In this, P will find out the prospect of getting YES in the attribute, and N will find out the prospect of getting NO. Put the values of P and N in the formula of entropy (class). Similar to this, calculate the entropy of each attribute of the table. Here it is taken the values of P_i and N_i by considering the values of that particular attribute with the corresponding (attribute) useful, which is entropy (class). Put those values in the entropy formula. After this, find out the gain from both the entropies, i.e., entropy (class) and entropy (attribute). After comparing with gain, the final decision tree is designed. For more correctness of the observed results, it is advised to calculate the accuracy, precision, and recall.

4.4.1 Entropy (Class)

Putting $P = 5$ & $N = 5$ in equation 4.1, we get Entropy = 1. For Table 4.1, the hacker recognition is the resultant attribute, which is our entropy by class from Eq. 4.1.

TABLE 4.1
Attributes for the Dataset for Decision Tree

S.No.	Metal	Plastic	Glass	Useful
1.	Reuse	Recycle	Melted	Yes
2.	Nonuse	Nonfunctional	Unmelted	No
3.	Reuse	Recycle	Melted	Yes
4.	Reuse	Recycle	Unmelted	Yes
5.	Nonuse	Recycle	Unmelted	No
6.	Reuse	Recycle	Melted	Yes
7.	Nonuse	Nonfunctional	Unmelted	No
8.	Nonuse	Nonfunctional	Unmelted	No
9.	Reuse	Recycle	Melted	Yes
10.	Nonuse	Recycle	Unmelted	No

4.4.1.1 For Metal

Using Equations 4.2 and 4.3, Entropy (Metal) =1

Hence, Gain= 1-1=0

Now, the first attribute of the dataset is taken, which is metal, to find out the value of P_i and N_i with the comparison of the entropy (class) from Equation 4.1 in useful attribute. First, the information gain (I) from Equation 4.2 and then the entropy of metal is calculated from Equation 4.3 using the above formula. Lastly, the gain of the attribute is found from Equation 4.4. Here, gain is zero represents the error-free calculation.

4.4.1.2 For Plastic

Entropy (Plastic)

The following data have been used for this purpose (Table 4.2).

Using Equations 4.2 and 4.3, E = 0.6041, Gain = 1 – 0.6041 = 0.395.

For plastic attribute, the information gain from Equation 4.2, entropy of the particular attribute, is calculated from Equation 4.3, and gain is calculated from Equation 4.4.

4.4.1.3 For Glass

The following data have been used for this purpose (Table 4.3).

TABLE 4.2
Data for Evaluation of Entropy of Plastic

Data/Probability	P_i	N_i	I
Recycle	5	2	0.863
Nonfunctional	0	3	0

TABLE 4.3
Data for Evaluation of Entropy of Glass

Data/Probability	P_i	N_i	I
Melted	4	0	0
Unmelted	1	5	0.650

It was found that, Entropy = 0.650 and Gain = 1- 0.39 = 0.61

For glass, the information gain (I) from Equation 4.2 and entropy is calculated from Equation 4.3. At last, final gain is calculated from Equation 4.4, which helped in the design of the decision tree. Hence, information gain of glass is found more than it is the node for formation of the decision tree, as given in Figure 4.2.

For attribute, glass entropy and gain are calculated, which is higher than 0.5 showing a suitable parameter for classification and bifurcation of decision tree formation. Further verification must be done by accuracy, precision, and recall. For calculation in accuracy, use Equation 4.5. In a classification experiment, the precision for a class is calculated as the number of true positives, divided by the total number of elements labelled as belonging to the positive class – Equation 4.6. Recall that this experiment can be defined as the number of true positives divided by the total number of factors that possess the positive class – Equation 4.7. Use data from Table 4.4.

$$\boldsymbol{Accuracy} = \frac{\text{No. of samples predicted correctly}}{\text{Total no. of Samples}} = \frac{5}{5+5} = 0.5 \quad (4.5)$$

$$\textbf{Precision} = \frac{TP}{TP+FP} = \frac{5}{5+5} = 0.5 \quad (4.6)$$

$$\boldsymbol{Recall} = \frac{TP}{TP+FP} = \frac{5}{5+0} = 1 \quad (4.7)$$

After making all the calculations, the final decision tree is designed, as shown in Figure 4.2.

Here, TP is truly positive and TN is a true negative. These are the observations that are correctly predicted positive and negative values, whereas FP and FN are

TABLE 4.4
Data for Evaluation of Entropy of Glass

Data/Probability	P_i	N_i	I (Info Gain)
Yes	5	5	1
No	0	0	0

false positives and false negatives. These values are observed when the actual class contradicts with the predictive class. Accuracy is calculated as the performance measure; it is the ratio of correctly predicted observation to the total observations. If accuracy is great, that means the model is good. Good accuracy is found only when symmetric datasets are there, and if any asymmetry is found, other measurements are there to calculate like precision, recall, etc. The above model showed 0.5 accuracies, 0.5 precision, and recall 1, which seems like a good classification of all the materials according to their properties. Hence, all the recyclable materials are classified in the smart dustbin and then further for processing for energy recycler and also supplied to industries for further reuse of materials. Those left out will further process other methods like shearing, magnetic, and eddy current until the last left out material that can decompose on land without effect to the environment. If any hazardous substance is left, then the incineration process is used for industries. Data classification by decision tree proves best in this case, and cloud source gives storage and security.

A good way is to go with green computing (Debnath 2015). Green computing mentions durable computing to the environment. This not only reduces the use of electrical power but also reduces the environmental waste when we are using a computer or laptop. This computing has the same achievement as green chemistry shows, which makes life easier and would make a life product more energy efficient. If this kind of 'GO GREEN' revolution is generated, the abandoned product and factory waste are more easily reproduced and can be biodegradable, less hazardous, and useful content for healthy living.

4.4.2 Ongoing Disposal Method of E-waste

The following techniques are used to eliminate e-wastes:

Landfills: If some of the materials end up in landfills, they may release toxins like mercury, nickel, copper, and cadmium. And if they reach land and water, they cause disaster for humans and animals. So, here we are trying to reduce these harmful effects by recycling plastics, etc. Avoid using plastic bags; instead, we can use jute bags. Rechargeable batteries are also a good option. Reduce junk mail, and try to manage food waste. Decompose organic remains that can be used as fertilizers instead of depositing them into landfills. Don't dispose of food immediately after consuming. Avoid buying unnecessary food; expiration dates on food products need to be checked. Check cooking utensils are good enough to cook food. They are also of alloy products with an ISI mark. Some of these measures can reduce landfills (Sivakumaran et al. 2017).

Acid baths: This procedure is utilized to extract copper from the circuit board in sulphuric acid (Sivakumaran et al. 2017). This is also a dangerous process; the worker should wear personal protection equipment (PPE) and follow the law. Use only scientific methods that align with the environment, health, and safety (EH&S) when disposing of an acid. Some acids are neutralized and poured into the sanitary sewer.

Hence, a highly efficient decision tree model is developed that can work in any city/smart city to decompose efficiently e-waste and protect the environment. And there is a need to develop more units for recycling of materials with highly efficient machines. Incineration is another waste treatment method, which includes the combustion of organic elements. These incinerator plants are the origin of dangerous pollutants. These plants need to follow the guidelines of proper waste discards. Most importantly, these incinerators require high-level machines to reduce hazardous pollutants.

4.5 CONCLUSION

It is recommended that waste prevention and minimization be the first step in every manufacturing unit. This will bring positive environmental outcomes. Reducing the amount of waste and toxicity is of the utmost importance. Manufacturing units need to design better quality products with long life spans; businesses will create new value-added offers for consumers. And it is the responsibility of consumers to change their consumption patterns by prioritizing sufficiency, sharing products, and applying product substitution so that all these products are reused by making product repairs, maintenance, and upgrades from time to time. Hence, the demand rises to rethink the product design for life longevity. It would also be better for the environment if not the whole product, but its components are recycled. If sometimes a product is not recycled and reused, then we can go for energy recovery, which will depend upon the type of waste constituent and the efficiency of energy recovery plants. Then, it is used for heat and power plants for producing electricity for industrial use. Hence, four big 'Rs' – reuse, recycle, repair, and research – are most important for saving life on the planet. It is to follow that all manufacturing organizations and consumers purchase good-quality appliances because they have transparency. Consumers need to be aware of all the bad effects of e-waste and not discard the tools to the same company or registered companies. Our little contribution will have a long impact on saving the environment. 'Be a technology planner, not a technology servant.' This chapter explains the importance of decision tree implementation in differentiating e-waste according to attributes without human interaction. This will create better human health while avoiding human interaction in rigorous methods. Now is the time to think smartly and make our present better. We have the responsibility to make our future safe for the next generation. The authors put an effort towards saving humanity through eco-friendly technology, even for flora and fauna.

REFERENCES

Bakas, I., Herczeg M., Blikravea, E. et al., 2016, Critical metals in discarded electronics, *Mapping recycling potentials from selected waste electronics in the Nordic region.*

Balde C.P. et al.2017. The Global E-Waste Monitor 2017: Quantities, Flows and Resources. United Nations University, International Telecommunication Union, and International Solid Waste Association.

Catania, V. and Ventura, D. 2014. An approach for monitoring and smart planning of urban solid waste management using smart -M3 platform, Computer Science, Proceedings of 15th conference of open innovations associations FRUCT.

Chen, A., Dietrich K., and Ho S., 2011, Developmental neurotoxicants in e-waste: An emerging health concern, *Environmental Health Perspectives* 119(4):431–438.

Debnath, B., Choudhuri, R. R., and Ghosh, S. K., 2015. E-waste management- A potential route to green computing, *International Conference on Solid Waste Management*, 3(6): 669–675.

Devika, S., 2010, *Environmental Impact of Improper Disposal of Electronic Waste,* IEEE, pp: 29–31.

Dubey, A., and Chouhan, U., 2012, Machine learning models to classify HIV membrane proteins, *Online Journal of Bioinformatics*, 13(2), 274–284.

Esmaeilian, B., Wang, B., Lewis, K., Duarte, F., Ratti, C., and Behdad, S., 2018, The future of waste management in smart and sustainable cities: A review and concept paper, *Waste Management* 81 (2018):177–195.

Gaidajis, G., Angelakoglou, K., and Aktsoglou D., 2010, E-waste: Environmental problems and current management, *Journal of Engineering Science and Technology.*

Garcia-Garcia, G., Stone, J., and Rahimifard, S., 2019, Opportunities for waste valorisation in the food industry e A case study with four UK food manufacturers, *Journal of Cleaner Production*, 211:1339–1356.

Han, J., and Kamber, M., 2013, *Data Mining: Concepts and Techniques*, Elsevier.

Hong, Y., and Valix, M, 2014, Bioleaching of electronic waste using acidophilic sulphur oxidizing bacteria, *Journal of Cleaner Production*, 65:465–472.

Lucier, Cristina A., and Gareau, Brian J., 2019, Electronic waste recycling and disposal: an overview, 15th FRUCT Conference Saint Petersburg, Russia. 10.5772/intechopen.85983

Miliute-Plepiene, J., and Yuhana, L., 2019, IVL Swedish Environmental Research Institute, 978-91-7883-090

Nyugen et al. 2018. Determinants of residents' E-waste recycling bahavioural Intention: A case study from Vietnam, *Sustainability*, 11(1): 64.

OSU, n.d., Safety instruction: Acid washing. Available from: https://ehs.oregonstate.edu/sites/ehs.oregonstate.edu/files/pdf/si/acid_bath_safety_si.pdf

Saxena, A., and Gautam, P., 2019, Emerging Trends in Cloud Computing: A Survey, International Conference on Efficacy of Software Tools for Mathematical Modeling ICESTMM'19, Organized by Thandomal Shahani Engineering College, Bandsra (w), Mumbai UGC Approved: 43602 Print ISSN 2349–5138.

Sivakumaran, R. P., Anandkumar Dr, K.M., and Shanmugasundaram, G., 2017, *Advances in Natural and Applied Sciences.* 11(8), 522–529.

Thi Thu Nguyen, H., Jay Hung, R., Hung Lee, C. and Thi Thu Nguyen, H. 2018, "Determinants of residents' e-waste recycling behavioral intention: A case study from Vietnam" *Sustainability*, 2019(11):164; 10.3390/su11010164.

Sivakumaran, 2013, E-waste management: Disposal and its impacts on the environment. *Universal Journal of Environmental Research and Technology*, 3(5): 531–537.

Xu, F., Wang, X., Sun, X., and Abdullah, A.T.M., 2014. Influencing factors and moderating factors of consumers' intentions to participate in e-waste recycling, 14(3): 2234–2240.

5 Possible Data Analytics Approaches for E-waste Management

Yumnam Jayanta Singh, Arup Chattopadhyay, and Sandip Chatterjee

5.1 INTRODUCTION

Proper handling of electronic waste (e-waste) in terms of collection, transport, treatment, and disposal is required to minimize environmental and health issues for all living beings. The current situation is alarming. There is rapid growth of the information and communication technology sector and the electronic manufacturing and assembly industry worldwide. There is also a high rate of obsolescence of the electronic products, resulting in uncontrolled accumulation of e-waste. Only about 17% of the total e-waste is collected and recycled globally. Over 27.7% of the population owns a smartphone. India only generates over 3.2 million metric tonnes of e-waste and related material per annum, ranking third in the world in e-waste generation after China and the USA. E-waste thus creates a universal crisis, leading to ecological degradation as well as serious health and environmental hazards. In most countries, unorganized, small, individual entrepreneurs as scrap dealers manage most of the e-waste. The main concern is lack of understanding among the various stakeholders about the ill consequence of the end-of-life products and the side effects. There is a directive restricting the use of certain hazardous substances in Waste Electrical and Electronic Equipment (WEEE). The European Union adopted the directive in 2003. The regulations of Reduction of Hazardous Substances (RoHS) set possible limits for substances such as polybrominated biphenyls, lead, cadmium, mercury, polybrominated diphenyl ethers, hexavalent chromium, etc. To stick to the compliances above, these substances should not be purposely added to manufactured goods. They should not go above the set permissible levels as a trace substance: 0.1% (1,000 ppm) for lead, hexavalent chromium, mercury, polybrominated biphenyls, polybrominated diphenyl ethers, etc. Different governments introduced penalties for non-compliance of their rules (Rodionov and Nakata 2011; Chatterjee 2012).

A key strategy to combat the hazards of e-waste while turning it into a business opportunity is to train the stakeholders through awareness programmes. The Ministry of Electronics and Information Technology (MeitY-Govt of India) is proactive in this regard and has implemented a series of pan-India awareness programmes and

 DOI: 10.1201/9781003317050-7

initiatives on e-waste management (Rodionov and Nakata 2011). Several self-learning materials are being produced using modern education technology tools like augmented reality and virtual reality. Similarly, e-waste management can be included in the syllabus of schools to teach the young generation the importance of the control of e-waste for a better future. In the same line, a short-term course on e-waste management is taught in institutes, and the Indian Institute of Technology Hyderabad (IITH) and Centre for Materials for Electronics Technology (C-MET), jointly started an M.Tech in E-waste Resource Engineering & Management (EREM), in the academic year 2020, with support by MeitY, Govt. of India. MeitY has already implemented a national-level 'Capacity Building of Government Employees on e-Waste.' Under this programme, through a 1-Day Training Programme, under 'Digital India initiative' through NIELIT Kolkata, many centres of NIELIT jointly imparted training to over 5,500 government officials at the PAN-India level in 29 states/UTs (NIELIT Kolkata 2020). The training was also imparted to officials of PSU, banks, industry, etc. Upon successfully completing the training, the candidates were awarded a joint 'Certificate' of NIELIT, MeitY, and MoEFCC. The many states of India adopted the E-Waste Management Rules, 2016. Accordingly, many state Pollution Control Boards (SPCBs) have started regular auditing of the Producer Responsibility Organizations (PROs), their collection centres, and the channels from e-waste collection centres to recyclers in their states as per the criteria provided by the Central Pollution Control Board (CPCB) to meet the rigid rules and guarantee that e-waste is being appropriately handled by the authorized recyclers (Chatterjee 2012).

This study proposed a data warehouse (DW) based novel data analytics approaches for e-waste management to produce more detailed and in-time management. The reports generated from this system will be useful for planning the smart city projects. It shows the technical steps to develop the system such as the process of extraction, transformation, and loading (ETL) of data and the collection of data from sources to produce final analytical results. It has many data analytics processes along with some related case studies such as (i) location locator of waste bins, status reporting of the filled level of the bins, (ii) support services for city municipality using remote sensing, WiFi, SMS, etc. to optimize the route to pick up the dumped waste, and (iii) re-clustering of the sites, etc., to help future waste predictions. It will help to examine current policies and frameworks for policymakers. Later, suitable tailor-made services can be developed that suit a particular city. Slowly the datasets and best success story can be added to improve the decision-making process. This study shows the possible use of an e-waste audit report to help the operator further examine and analyze the report to the level of spatial locations or sectors. It reports the achievement of a specific training when the colour scheme is used in the traffic control system. The operator can make a plan to finish training in the 'Red Zone' city area before the 'Green Zone' areas from the pictorial report. During any city planning, the related information can be accessed from the system to prepare the possible action plan and set an appropriate ground-level plan for the real action as per the priorities. It will help the authority to plan the critical activities and their schedules. The study can help compare the cities of a state with respect to the e-waste management process and the associated services. Another important facility of the study is to predict future requirements. The periodic audit data can help the authority recheck the rules, and modify action plans to issue necessary revised directives from time to time. The study also proposes a dashboard facility to show at-a-glance views of summary performance indicators. The user can download the report as per demand and can correlate the performances in their respective sections at field level. It will help in estimation of the amount of reusable, recyclable items for both commercial purposes as well as mitigation of e-waste management issues. The e-waste data can also be linked to the individual products so that usages of hazardous elements may be minimized or restricted as per the RoHS guidelines. The authority then may implement the revised model and plan for the next set of actions. Thus, the study may also be used for attaining a better business model for any startup or entrepreneurial initiatives of local unemployed youths.

5.2 EARLIER STUDIES

Several studies are attempted in similar areas of waste management across the globe (Zurbrügg, Caniato and Vaccari 2014; Khan and Samadder 2014). It is found that one of the commonly used processes is land disposal. The GIS facilities are used to optimize the routes for collecting solid wastes to minimize the cost. Fuzzy decision-making scheme is applied to identify the dumping sites. Some of the main effects of land disposal are pollution, health hazards, and a threat to the environment. It is also suggested for a user-oriented waste management system where users from different levels and sections can contribute in proper management of the e-waste. Usage of a standardized approach and multiple case studies are also essential to develop a sustainable model. A survey of possible integrated waste management solutions is provided (Hornsby et al. 2017). The consultancy services of the stakeholder are used to bring a better-integrated waste management system. It considers the viability, socio-economic, and ecological issues. Surveys were conducted to understand the sentiments of the public. The waste management also uses the decision support frameworks to bring a future sustainable model (Karmperis et al. 2013). It assesses the life cycle of the waste. Game theory was used for modelling and analyzing a scenario before developing a better model. It helps in developing a decision support system for the business. The fuzzy approach is used to model the uncertainty situations (Lolli et al. 2016). More stakeholder participation in the study of the environment will result in better understanding of the real need and priorities. So, the fuzzy approaches are used to model the uncertainty situation or the views. Fuzzy approaches are useful for developing a more reliable system.

The concept of weight factors is used in the analysis of the municipal solid waste management during urbanization (Zhou et al. 2019). An analytical hierarchical and multi-criteria decision process is used for solid waste treatment (Antonopoulos et al. 2014). The performance-based evaluation is conducted on contributing parameters (either ecological or social or economic), and it analyses the rank of the performances. It suggests for a biological aerobic handling process of the waste materials. Assessment of waste management is executed using the analytic hierarchy process (Tsydenova, Vázquez Morillas and Cruz Salas 2018). This facility helps to break a large problem into its small elements for better comparison. It allows decision making based on the priorities. A study shows a specific need for waste management regulations related to medical issues (Aung, Luan and Xu 2019). It studies waste management related to health care services as per the World Health Organization norms. Several questioners and customer-based data are collected for the study. The multi-criteria-based decision support system helps to model the success criteria. In a similar line, many studies aim to develop some expert system for the waste management system. There is a need to enhance public participation in the same concern (Higgs 2006). Several data are collected from the stakeholder and are validated to develop stakeholder-centric management. The suggestion is provided on the risk analysis and cost-benefit analysis for the waste management project (Karmperis et al. 2012). Different versions of decision support are modelled, aiming for a sustainable model. The risk analyses are taken as one of the main contributing parameters for the study. The risk report suggests choosing the less risk-related parameters.

It is essential to the modelled sustainable process that the waste's life cycle is assessed to bring an economic model (Emery et al. 2007). The life cycles of most of the waste are different. It will not be sustainable to aim for a particle or item-based waste management system. Integrated waste management is suggested for efficient management and controlling its associated risks and effects. The use of GIS facilities optimized the travel route during waste collection. The waste bin network was developed to minimize the travel by pickup trucks (Chalkias and Lasaridi 2009; Tavares et al. 2009). The cloud or WiFi enabled pickup truck can understand and follow the shortest route. This helps in minimizing the cost through optimization of consumption of the fuels and life span of the pickup trucks. Many studies are conducted to suit a region or solve a particular problem. An interconnected system is required to properly handle the global e-waste and convert to a potential new profession for the startups (Münster and Meibom 2011). Most of the proposed studies aim to solve a specific local problem. The data are scattered in their specific domain. Waste management is a global problem. An integrated global e-waste management system is required, where the best practises can be shared between the different countries. Every country's youths can be brought together in a single platform to exchange success stories and learn from each other.

5.3 RESEARCH GAP

Artificial intelligence techniques, data warehouse, and data analytics are applied towards better management and future planning. Many of the studies concentrate on controlling the waste or the e-waste. Some systems are required that can evaluate existing situations and predict future requirements. Already the globe is facing a severe crisis because of the ever-increasing accumulation of e-waste materials. The scenario across the world has worsened from the protective materials of the Covid-19 pandemic. Simultaneously, very few effective measures have been taken to convert this waste management to a profitable business so that interested individuals may be attracted towards e-waste management as a value-added profession.

Most of the studies use several decentralized assessment tools, and data are also located loosely. A user-oriented waste management system is missing where users from different levels can jointly contribute and work together. Many facilities, such as a set of rules and the process of recycling or waste management, are available. However, a proper linking between such processes and all the stakeholders is very much required. It needs suitable tailor-made services that should be usable and suitable for a particular city. Studies of literature reflect that e-waste management may become a new profession for startups. However, useful data and the methodology are not visible for a sustainable model.

There is a need for periodic study of the e-waste audit to see the acceptance level of the rules from different authorities and control the training providers' contents. Several reports or indications are required in visually comprehensible colour format so that the user may analyze the reports and make effective management decisions. The priority scheduling of the services is necessary so that some modelling can be performed before a real action to assess requirements and also arrange the working schedule. It is recommended to have a prediction system for any future needs to predict numbers of required recycling units and their capacities. The periodic audit data can help identify the list of agencies that do not follow the rules and the necessary directive. The study can link the concerned department, such as departments of health and environment, to join hands to share the available welfare schemes. Further, a dashboard facility is suggested to enable a graphical user interface to show at a glance the summary of performances. The facility for delivering a progress report of the department or the service is also necessary. This may attract many new users, and even authority can monitor the profiles of visitors. Many dashboard items are useful for measuring the efficiencies of the services. Most of the reports are represented in visual format for easy interpretation of the users. The individuals can become more aware of the available services from childhood. Thus, a study is required to find a better business model with a multi-disciplinary approach for converting e-waste management to a new profession for startups or individuals.

5.4 STUDY METHODOLOGY

This study on 'possible data analytics approaches' aims to answer some business questions to manage the waste materials better. It starts by collecting related input data, cleaning the data, and finding out the relevant data. It aims to find some useful patterns. Then, possible business models are prepared. The concerned authorities learn through systematic observation and interpretation of the results, and the new requirement is incorporated in decision making. The proposed data analytics system has several subcomponents; (a) data sources, (b) ETL process, (c) data warehouse, (d) data mart, (e) integration module, and (f) analytical task operations. Theories for setting up a data warehouse are learned from earlier studies (Singh, Singh and Singh 2020). To supply data to the data sources, the data are collected from concerned sources. The detail data are stored in a particular table called a dimension table and in the cube format to enable modelling and view in many dimensions. Another unique table, called a fact table, is used to connect the dimension tables and measure the consolidated statistical data. A star schema is used to connect the fact table to the dimension tables. As source data, many related data are also extracted through the existing MIS system. Data sources are gathered in government agencies and rules, infrastructure details, lists of trainers and auditors, etc. The data are also grouped as per appropriate subgroups of stakeholders, such as the list of the manufacturer, product producer, consumer (individual or bulk), collection centres, product dealers, e-retailer, refurbisher, dismantler, recycler, trainers, auditors and other related non-government organizations (NGO), etc.

5.4.1 The Data Sources

Normally the required input data are collected from many different trustworthy sources. They can be extracted from some existing data warehouse system also. Possible error handling techniques can be applied in case there are errors in the data. Some of the error handling tips are available in earlier studies (Singh et al. 2019). The error-free data are stored in the possible cube format. This enables the user to model the data in many dimensions, offering scope for visualization. This helps the operator or the users to understand the report in pictorial form in a simpler way. This study used different types of inputs such as the 'set of rules' given by authorities, the list of the agencies that approve the firms, the e-waste audit providers, training providers, etc. There are other types of input such as the relevant information about the firms that generate e-waste, a list of their customers, etc. Another set of input data may be the data related to spatial location of the dumping sites, major waste bins, collection centres, recyclers, etc. Some more specific input data and their attributes are given in Table 5.1.

5.4.2 The Extraction, Transformation, and Loading (ETL) Process

ETL is the design process of any data warehouse system. As the name suggests, it means the extraction, transformation, and loading of the data. During 'extraction,' the data from different sources are extracted from other data formats (the used tools or the software), units (may vary from country to country and also among stakeholders), and types (structure or un-structure). During 'transformation,' the data are

TABLE 5.1
Sample Input Data or Related Attributes

Govt. Agency and Rules	Agency or Service or Private Related Provider
The agency that approves the firms	Firms that generate e-waste, their customers
E-waste audit providers	Firms that are entitled to the e-waste compliance audit
E-waste training providers	Private training providers
List of the applicable rules	Public awareness programmes
Location of major waste bins	Types of public complaints
Agency list related to e-waste, etc.	List of NGOs who work related to e-waste, etc.

transformed into a better way per some defined rules. The rules could be syntactic or semantic. During this process, the system also handles better cleaning, missing data, outliers, etc. During the 'loading' process, the error-free data are loaded to the data warehouse system. Appropriate timing for loading, backups, and refreshing times are followed. These timings are very important while dealing with a live warehouse system with lots of live users.

5.4.3 The Warehouse System

A simple data warehouse system is used to store the integrated data. It has a connection to many different types of data from many data sources. Such a design enables the strategic flexibility towards future business planning of any firms and also in making decisions. It connects to all the given subcomponents. A sample data warehouse-based analytical system for e-waste management is shown in Figure 5.1. All the possible subcomponents or modules are integrated to form a complete system.

As mentioned in the introduction section, some of the direct data can be collected from report of the 'Capacity Building on E-waste,' where NIELIT trained over 5,500 government officials at the PAN-India level in most of the Indian states. It gives various applicable E-Waste Management Rules, 2016, provided by the SPCBs and the rules of CPCB for the collection centres and recyclers. Such more typical data attributes are given in Table 5.1. Usually, a data warehouse consists of most of the cleaned data. The erroneous or unclean data from the above sources are processed for cleaning based on the ETL process and designing a data warehouse system (Powell 2011).

5.4.4 The Data Mart

Data mart theories are used to group similar and related types of services for better management. They will enable users to process similar kinds of data or services, easily back them up, and provide an easy alteration for business changes. Some sample marts are the sets of rules, the training providers, public awareness

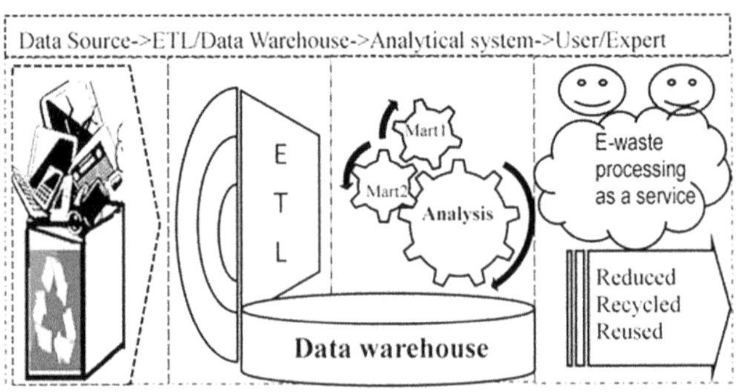

FIGURE 5.1 Sample data warehouse-based analytical system for e-waste management.

TABLE 5.2
Sample of Potential Data Mart Attributes and Some Possible Analytics Processes

Attribute of Marts	Possible Analytics Processes
Sets of rules	List of approved e-waste recyclers
Training providers	Frequently raised customer complaints
Public awareness programmes	Training providers and their coverage
Colour-based labelling of regions	District list and their status of e-waste
District-wise status	Highest used or least used services
Customer complaints	Most compliance and complaint rules
Concerned Govt. department and officers	Periodic status of the training provided
MIS or tools used by City Municipality	The complaint related to health issues, etc.
	The capacity of the recyclers in a state
Network map of the major waste bins	Any related operations
List of e-waste compliance auditors	
List of recyclers	
Sets of analytical processes for officers	

programmes, colour-coded labelling of regions, the district-wise status, customer complaints, concerned government departments and officers, MIS or available tools used by the City Municipality, a network of the major waste bins (if any), a list of e-waste auditors, a list of recyclers, etc., and the sets of analytical processes for the owner of this system, etc. Table 5.2 shows some potential attributes of the data mart and some possible analytics processes for proper management.

5.4.5 Analytical Task Operations

The study proposed various analytics processes to provide solutions to related business requirements. Some of the possible examples are given in the following sections. Such analytical techniques will help authorities in future planning. Finally, it will enable to convert such a study as 'waste processing as a service.' This concept can be considered in the same line as cloud facilities as services on 'payment' mode. Doing so will help the waste/e-waste management authority and the respective states attract, establish, and retain several 'startups' or 'individual entrepreneurs' to support the government to handle the waste jointly while adequately earning a livelihood from the services provided. The citizen also can avail extra services such as call a pickup truck, request special training or an awareness programme for their societies, or even join the related businesses. Currently, most of the waste management services are 'supply-driven' affairs, but slowly it can be converted to 'demand-driven' activities. In the same way, analytical processes can be applied to many sectors for better productivity. Doing so will help in generating lots of expertise or jobs for experts and will open a new vertical for services with job opportunities.

The owner or the users of this data analytics system have been provided with several services to process as required by the business. The user can find out the list of all the registered electronic goods manufacturer firms that may produce the e-waste in a particular state. The expected total amount of annual e-waste can be calculated using a simple survey. It can also find the list of approved e-waste recyclers. The study can drill down to each district level, and their status of e-waste can be extracted from those dimension tables and their data in the cube format. Thus, we can know the location or region's status either under the tolerable limit or out of control, etc. The frequent rise in customer complaints and related health issues, etc., can be extracted from the system. The system will help determine the list of the related training providers, their coverage, and the next requirement. The data can be linked between total or possible number of concerns, firms, and compliance reports. This will help find out the need to set up more e-waste auditors or trainers. The feedback collected from them will reflect on the highest used or least used services and rules. Accordingly, a state's authority will come to know about the difference of capacity between the generated amount of e-waste and the handling capacity of the recyclers and their concerns.

Similarly, many related analytical operations may be performed. Time to time, several prediction services can be processed to assist the authority to design an ecosystem of 'reduced, recycled, and reused.' However, the correctness of the system will depend on the data collected and populated into the above analytical system.

5.5 A CASE STUDY: A DATA ANALYTICS APPROACH FOR EFFICIENT E-WASTE MANAGEMENT FOR A CITY

The case study considered in this study is based on a mini prototype that may be used to 'efficiently manage various types of e-waste for an Indian city.' Such a system can be represented in Figure 5.2. All the above best services are combined and called 'waste processing as a service.' It connects the concerned authority, the rule or the available services, and the demands from the public. This e-waste management system may have the following components (a) municipality services,

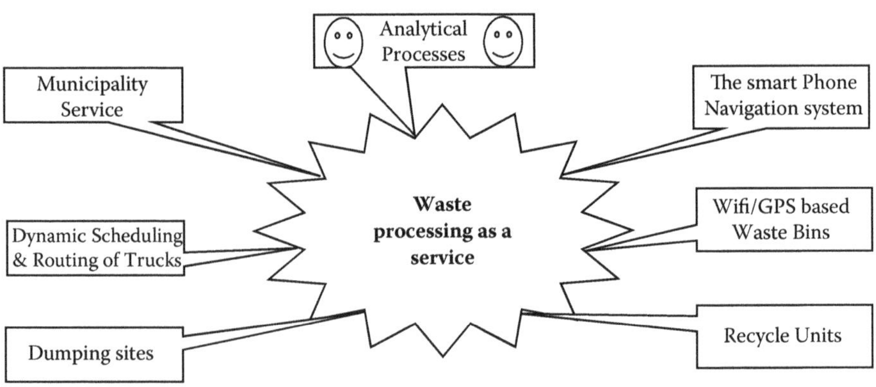

FIGURE 5.2 Waste processing as a service for e-waste management for a city.

(b) dynamic scheduling and routing of the pickup trucks, (c) dumping sites, (d) smartphone navigation system for officers and the public, (e) WiFi or GPS-based waste bins, (f) recycle units, and g) overall analytical processes.

5.5.1 Municipality Services

A municipality or an assigned authority can have several waste management related services; training of the staff or the public, the audit of the activities, monitoring services, managing the pickup truck and bins, etc. With the rise of smart cities, the responsibilities of the municipalities increase. Many electronic scraps are sold at the best price in India. Then, with existing best practises, which are often unscientific and informal, the e-waste is recycled and reused. Municipality plays a vital role in regulating the organized or unorganized sector. It helps in generating jobs as a viable business for youths.

5.5.2 Dynamic Scheduling and Routing of Pickup Trucks

Daily, many vehicles are used for the pickup of waste materials. Such vehicles can cause pollution from the payload they carry, which can affect the health and the environment. At the same time, saving the lifetime of the vehicle, the fuel, and workforce involved are also crucial for the country's economy. The waste management business is a regular activity and will always continue, so its cost should be controlled, optimized, and ideally converted to an income-based business opportunity. More information is provided in the next section.

5.5.3 Dumping Sites

The information on the dumping site is updated regularly. Data analytics processes are performed to understand the capacity of a site or entire available sites and the staff requirement. This information will lead the authority to resign or allot new spaces. From the available data, the authority can predict the future requirement and future course of actions.

5.5.4 Smart Phone, Navigation System for the Officers and Public

Some of the online services of the system are provided to the public. The public can request a special pickup facility on a payment basis. The interested public group may ask for arranging the training or awareness programmes. The officer can use the provided services for their daily assignment from anywhere. The authority can manage the assigned staff and also analyze the responses or the feedback of the public.

5.5.5 WiFi or GPS-Based Waste Bins

The waste bins are proposed to enable WiFi and also with GPS. The sensing can be periodic or per demand. This will allow communication between the navigation system, the authority, and the pickup truck. The study recommends solar-based chargeable devices.

5.5.6 Recycle Units

The recycle units can be owned by a government agency or authorized private agencies. The recycling units' data are collected for the dumping site to understand the types of activities and capacities. The generated data are used to understand the varieties of waste, which will help in planning of recycle and reuse processes.

5.5.7 Overall Analytical Processes

All the above components are integrated under this unit as the system's heart enables the collective service of 'waste was processing as a service.' These units connect to many different components and their available data. Most of the business logic and model are executed using this unit by the various users or the authorities.

5.5.8 Optimization of Collection Routes Using WiFi-Based Smart Bins

Figure 5.3 shows the proximity sensors and WiFi-based 'smart bins.' A state or concerned authority can develop its infrastructure such as a concerned department, waste dumping sites, waste bins, pickup trucks, dismantling or recycling units, and navigation systems to connect the facilities. It also uses the possible networking facilities such as WiFi, remote sensing, and proximity sensors. The 'proximity sensors' are attached to the waste bins to make them 'smart bins,' to get the filled level of the waste that day. The bin's statuses are digitally reported to the system operator or the drivers of the pickup trucks. The pickup truck uses the data to decide the optimized route to reach the 'waste bins.' A small status table can be used for considering the filled percentage or status of a bin, and corresponding 'action' can be assigned and executed. Thus, the authority can decide accordingly based on the

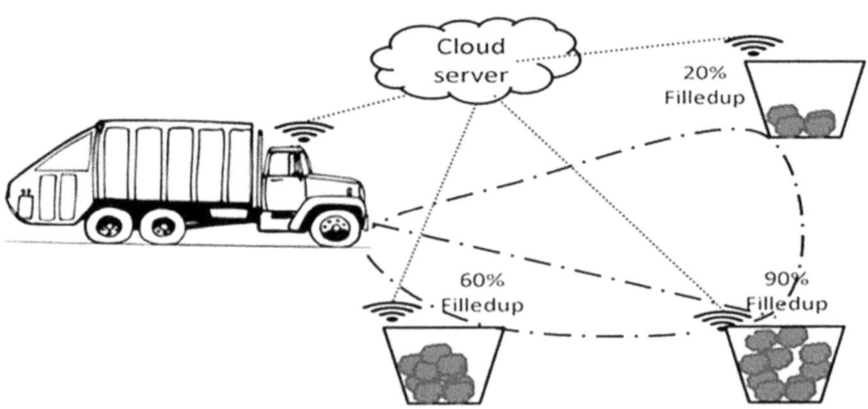

FIGURE 5.3 The scenario of a WiFi-based pickup truck and the 'smart bins'.

bins' filled level for the relocation of the bins. For example, if the waste bins are filled up faster than the average expected days, more bins may be provided. In another case, if the bins took longer to fill up, then they may require alternate plans like relocation to a more populated region. The proximity sensors will help optimize the pickup routes or re-clustering of the sites, etc., to support future waste predictions. The similar process or activity is used for considering appropriate dumping sites of the waste. It will avoid driving a pickup truck to an overloaded site. Thus, optimized routes can be generated, and future planning can be done based on the daily report generated and best practices that will minimize fuel consumption. In the same way, optimized networks or routes can be modelled showing the shortest routes among the 'bins' and dumping site.

5.6 DATA ANALYTICS APPROACHES AND RESULT DISCUSSIONS

There are different levels of users for such a system-level authority: middle level, operator level, individual users, etc. Various analytics processes are used to get the solution for related business requirements of the authority and future planning, such as the status of the cities concerning the periodic auditing and the status comparison of the cities or the regions. Accurate data are provided from many sources during the system's setup and by the corresponding operation during execution of the service. The integrated and summarized data are stored inside the corresponding data marts while connecting to all the source components. Most of the analytical operations are performed on these data marts.

5.6.1 Study on the E-waste Audit

An e-waste audit has been mandated by CPCB for all states/UTs; the different state-level authorities (SPCBs) are in the process of implementing the audit at the field level. These audits are carried out under strict norms, rules, and formats for all stakeholders concerning e-waste management. This requires a better understanding of the e-waste management rules and their implications in both manufacturing and disposal streams. Consequently, training providers are also coming up under government initiatives for better public awareness. A different audit can be conducted to ensure services. Some possible audit areas are compliance of sets of mandatory rules, district-wise status, lists of e-waste compliance auditors, frequently raised customer complaints, training providers, commonly used rules and least used services, complaints related to health issues, etc. Figure 5.4 shows the analysis report of the status of the audit reports. The report indicates that currently, the audit status is 20%, and the remaining 80% is pending and yet to perform. Further, the operator can drill down the report to the level of locations or sectors (health, environment, etc.).

5.6.2 Study on Achievement Report of Specific Training

Consider that there are 30 cities so far registered for this study, and ten cities belong to three different statuses. This colour scheme of green, yellow, and red

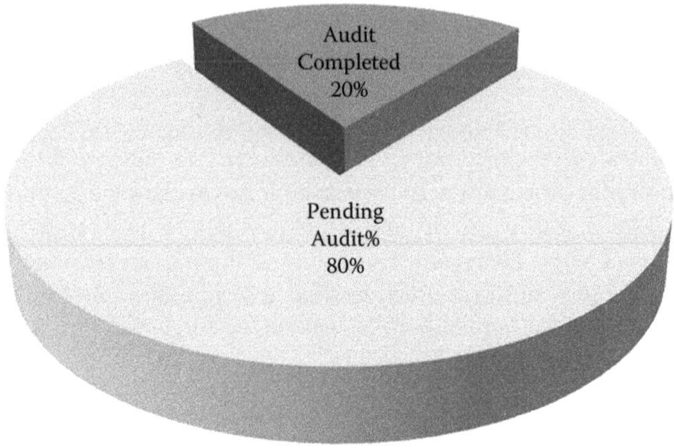

FIGURE 5.4 Sample e-waste audit report of a state.

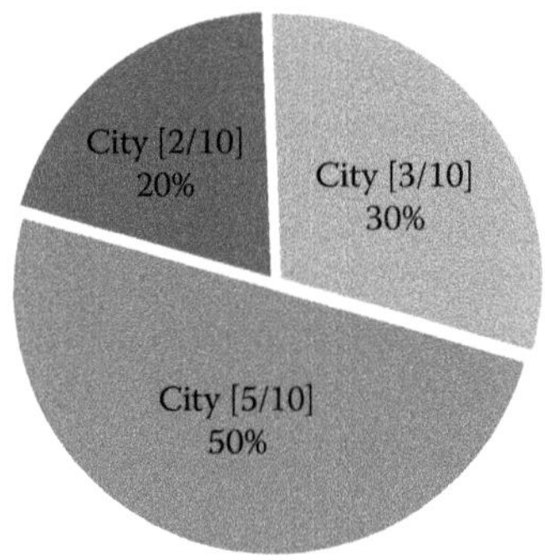

FIGURE 5.5 Sample of achievement report on training of some selected cities.

represents the severity levels used in the traffic system. The green colour may represent lesser severity cities with less e-waste, whereas the red colour may represent the cities that need immediate attention. Figure 5.5 shows the simplified sample analysis report of the cities in the colour marking scheme. It shows the number of cities that lie in each of the colour levels concerning the selected cities list. The operator may appropriately implement the training programme, for example, by starting with the 'red zone' city area before the 'green zone' areas from such a pictorial report.

5.6.3 Study on the Status of a City during the Planning Phase

The data mart contains summarized data from different sources. During any city planning, the related information can be retrieved from the system. Consider a list of activities to perform such as conducting an awareness programme for a set of individuals or staff, the staff in the collection unit, and a dumping site. The city's current status concerning the set of chosen actions can be analyzed before the real action takes place considering the priorities (if any). Correspondingly,

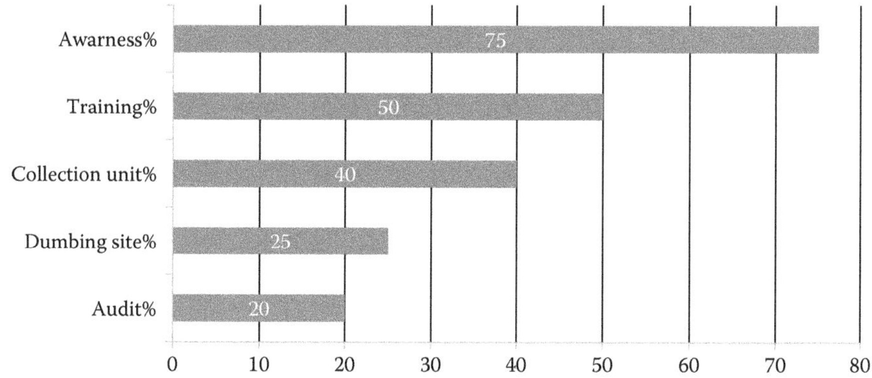

	Audit%	Dumbing site%	Collection unit%	Training%	Awarness%
Series1	20	25	40	50	75

FIGURE 5.6 Sample analysis report of an e-waste system of a state (for planning).

the required material can be gathered, the staff can be mobilized, and the working schedule can be prepared. Several earlier data can also be retrieved to help compare the performance based on the daily, monthly, or yearly basis and support in better planning. Figure 5.6 shows the analysis report of an e-waste system of a state at a planning phase. Here, the authority can plan the activity to give priority to their schedules.

5.6.4 Study on the Comparison of the Cities

The study can help compare the cities of a state with respect to the performance in the e-waste process and its associated services. Cities may be compared on services provided such as awareness, departmental training, and executed numbers of the audit. Awareness or training is an essential activity for all stakeholder. After an audit, the authority will clearly understand a situation and plan its next action for improvement. It often happens that even though training is provided, not enough audits are carried out, leading to unverified or unchecked data. Figure 5.7 shows the report of the status of each city on the provided e-waste services. In such a way, 'e-waste processing' can be used as a service to better the entire population.

5.6.5 Prediction of Future Requirements

The study can predict numbers of dumping sites or requirements for augmentation of an e-waste dismantling/recycling facility for next year based on available current data. It can also predict requirements of new technology to be introduced in the recycle units and their capacities. The periodic audit data can help find the list of agencies that do not follow the rules, and the necessary directive can be issued from time to time. If needed, extra hands-on training may be provided.

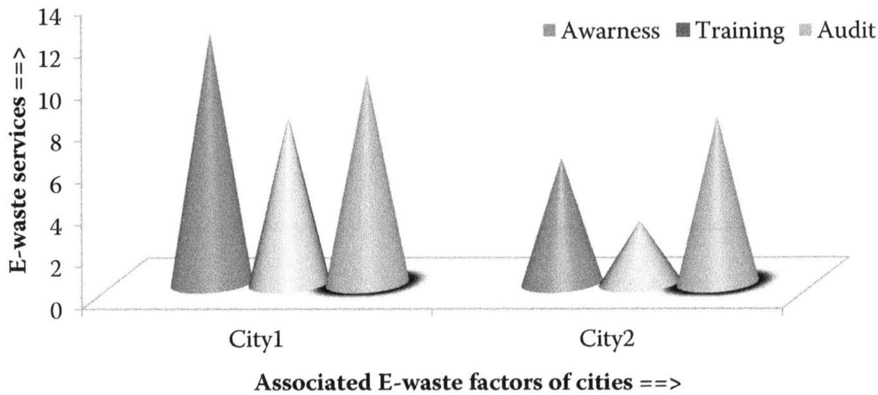

FIGURE 5.7 Sample analysis report comparison of the cities of a state.

5.6.6 Study on Health and Environment-Related Data

There are many issues related to health and the environment concerning e-waste management. Many times, the residue of the extracted waste material is not disposed of properly. Domestic animals like cows, buffalos, and goats may consume the same, and the dairy products may be highly contaminated and harmful. For instance, contaminated milk is hazardous for human consumption. The study can help set up and monitor such activities to minimize the impact on the human lifestyle, health services, and the environment and make our planet a better place for living. The concerned department, such as health and the environment, may be associated with joining hands to share their respective available welfare schemes.

5.6.7 Dashboard Facility

A dashboard facility with a graphical user interface is absolutely essential to show the at-a-glance views of the services' summary performance indicators as a progress report of the department or the service. It can attract many new users, and also authorities can monitor the visitors of the same. Many dashboard items are useful for measuring the efficiencies of the services. Most of the reports are represented in a visual format for easy usage by the users. The user can download the report as per demand and easily correlate the performance for transference purposes. The customer can also provide feedback or lodge a complaint. A well-designed dashboard works as a campaign and can attract lots of business supporting any service's digital marketing.

5.6.8 Study on Better Earning Business Model

The model can be developed to see the relation between the collection amount of the e-waste and the available recycler's capacity. It will help to roughly calculate the amount of reusable, recyclable items for business purposes. The e-waste data can be linked to the product so that some usages of hazardous products can be minimized or modified. Then, the authority can plan for next actions. Many possible schemes can be introduced to attract more partners for better management of the e-waste. The individual can be given more awareness to make use of the available services and sensitize the children at schools as they are the future citizens.

5.7 CONCLUSION

This chapter proposed a data warehouse-based e-waste management system that will help achieve a more comprehensive resource and time management. The reports generated from this system will be useful for city planning and future business. The method may connect several services. It can drill down the information to a specific level of a particular waste bin or a concerned staff while also helping authorities with valuable insights at the macro level for resource mobilization for the next actions. The data analytics processes enable us to track resources, like finding the location of waste bins, checking the filled level of bins, optimizing the pickup routes or re-clustering of the sites, etc., and thus helping future waste predictions. This study on possible analytics approaches may help all the concerned stakeholders like local municipalities, CPCBs, SPCBs, producers, consumers, bulk consumers, collection centres, refurbishers, dismantlers, recyclers, NGOs, etc., in better management of e-waste. To create user-level skills, several self-learning materials can be produced using augmented reality and virtual reality for the target audience. It will help the cities solve the imminent environmental and public health challenges associated with waste generation and inadequate waste collection, of which e-waste, its transportation, treatment, and disposal is a substantial portion. It will enable the policymakers and entrepreneurs to examine the current situation, policies, and frameworks while aiming for a manageable and productive future. Further, the datasets and best success stories can be progressively added towards a more robust decision support system and a better future for everyone. In line with the prevalent value-added cloud computing services, available for different business purposes, the proposed 'e-waste processing' services can also be modelled as a paid service. Tailor-made services can be developed and offered for a specific city, resulting in a business vertical based on a specific scenario and requirements.

REFERENCES

Antonopoulos, I. S., Perkoulidis, G., Logothetis, D. and Karkanias, C., 2014. Ranking municipal solid waste treatment alternatives considering sustainability criteria using the analytical hierarchical process tool. *Resour Conserv Recycl*, 86, pp.149–159.

Aung, T. S., Luan, S. and Xu, Q., 2019. Application of multi criteria decision approach for the analysis of medical waste management systems in Myanmar. *J Clean Prod*, 222, pp.733–745.

Chalkias, C. and Lasaridi, K., 2009. A GIS-based model for the optimization of municipal solid waste collection: The case study of Nikea, Athens, Greece. *WSEAS Trans Environ Dev*, pp.11–15.

Chatterjee, S., 2012. India's readiness on ROHS directives: A strategic analysis. *GJSFR*, 10(1), pp.14–26.

Chatterjee, S., 2012. Sustainable electronic waste management & recycling process. *Am J Environ Sci*, 2(1), pp.23–33.

Emery, A., Davies, A., Griffiths, A. and Williams, K., 2007. Environmental and economic modelling: A case study of municipal solid waste management scenarios in Wales. *Resour Conserv Recycl*, 49(3), pp.244–263.

Higgs, G., 2006. Integrating multi criteria techniques with geographical information systems in waste facility location to enhance public participation. *Waste Manag Res*, 24(2), pp.105–117.

Hornsby, C., Ripa, M., Vassillo, C. and Ulgiati, S., 2017. A roadmap towards integrated assessment and participatory strategies in support of decision-making processes. The case of urban waste management. *J Clean Prod*, 142, pp.157–172.

Karmperis, A.C., Sotirchos, A., Aravossis, K. and Tatsiopoulos, I.P., 2012. Waste management project's alternatives: A risk-based multi-criteria assessment approach. *J Waste Manag*, 32(1), pp.194–212.

Karmperis, A.C., Aravossis, K., Tatsiopoulos, I.P. and Sotirchos, A., 2013. Decision support models for solid waste management: Review and game-theoretic approaches. *J Waste Manag*, 33(5), pp.1290–1301.

Khan, D. and Samadder, S.R., 2014. Municipal solid waste management using GIS aided methods: A mini-review. *Waste Manag Res*, 32(11), pp.1049–1062.

Lolli, F., Ishizaka, A., Gamberini, R., Rimini, B., Ferrari, A.M., Marinelli, S. and Savazza, R., 2016. Waste treatment: An environmental, economic and social analysis with a new group fuzzy PROMETHEE approach. *Clean Technol Environ Policy*, 18(5), pp.1317–1332.

Münster, M. and Meibom, P., 2011. Optimization of the use of waste in the future energy system. *Energy*, 36(3), pp.1612–1622.

Powell, G. J., 2011. *Oracle data warehouse tuning for 10g*. Elsevier.

Rodionov, M. and Nakata, T., 2011. Design of an optimal waste utilization system: a case study in St. Petersburg, Russia. *Sustainability*, 3(9), pp.1486–1509.

Singh, Y. S., Singh, Y. K., Devi, N. S. and Singh, Y. J., 2019. Easy designing steps of a local data warehouse for possible analytical data processing. *ADBU J Eng Technol*, 8.

Singh, Y. S., Singh, Y. K. and Singh, Y. J., 2020. Local analytical system for early epidemic detection, in COVID-19: Prediction, decision-making, and its impacts. *LNDECT*, 60.

Tavares, G., Zsigraiova, Z., Semiao, V. and Carvalho, M. D. G., 2009. 'Optimisation of MSW collection routes for minimum fuel consumption using 3D GIS modelling. *J Waste Manag*, 29(3), pp.1176–1185.

Tsydenova, N., Vázquez Morillas, A. and Cruz Salas, A. A., 2018. Sustainability assessment of waste management system for Mexico City (Mexico)-Based on the analytic hierarchy process. *Recycling*, 3(3), p.45.

Zhou, Z., Chi, Y., Dong, J., Tang, Y. and Ni, M., 2019. Model development of sustainability assessment from a life cycle perspective: A case study on China's waste management systems. *J Clean Prod*, 210, pp.1005–1014.

Zurbrügg, C., Caniato, M. and Vaccari, M., 2014. How assessment methods can support solid waste management in developing countries-a critical review. *Sustainability*, 6(2), pp.545–570.

6 Sustainability Aspects of E-waste Supply Chain Network Role of New and Emerging Technologies

Biswajit Debnath, Rohit Roy Chowdhury, Chayan Acharya, and Arnab Guha

6.1 INTRODUCTION

Electronic devices like smartphones, televisions, laptops, sound systems, and other devices are extremely important to modern culture. The main factors contributing to the shorter lifetime of electronic devices include technological development, rapid invention, rising customer demand, and fashionable but irreparable design of the gadgets. From 44.7 million metric tonnes in 2016 to 52 million metric tonnes in 2019, the amount of electronic waste (e-waste) generation worldwide increased exponentially (Balde et al. 2017; Forti et al. 2020). Only 17.4% of the e-waste was recycled in 2019 worldwide (Forti et al. 2020). Hence, it is essential to develop an efficient e-waste management system for better management of e-waste.

E-waste management is a complex area that requires an efficient supply chain network (SCN) and smooth operations management (Doan et al. 2019; Wibowo et al. 2021). The SCN of e-waste is complex as well as interesting (Debnath 2020). A generic e-waste SCN is shown in Figure 6.1. From the global scenario, there are multiple stakeholders along the supply chain of e-waste. It includes the informal sector, semi-informal sector, formal recyclers, dismantlers, third-party recyclers, etc. (Debnath 2022). There are certain differences, which vary from nation to nation. In developing nations, such as India, the e-waste supply chain is overruled by the informal sector due to lack of awareness (Dwivedy and Mittal 2010). The trans-boundary shipment of e-waste is also boosting the informal sector in spite of the existence of legislation (UNEP 2013). It is a well-known fact that e-waste is shipped to countries like India, Bangladesh, China, Nigeria, Vietnam, etc., for recycling. This is known as trans-boundary movement of e-waste, which is a major problem (Dias et al. 2022a). Often, a large portion of this imported e-waste, primarily the PCBs, is handled by the informal sector, who employ rudimentary

DOI: 10.1201/9781003317050-8

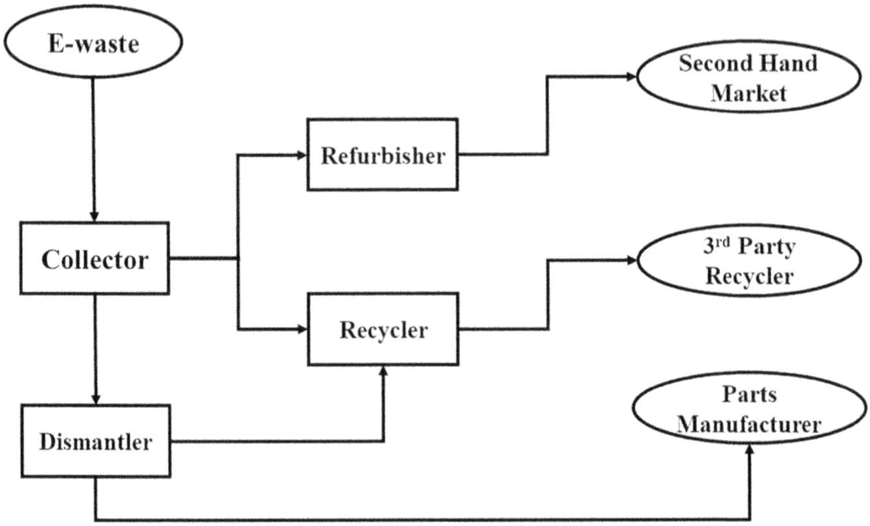

FIGURE 6.1 Generic e-waste supply chain network.

techniques to recover metals. As a result, environmentally friendly management of e-waste becomes an issue (Dias et al. 2022b). Basel Action Network (BAN) has put an injunction on exporting and importing e-waste from other countries. However, the loopholes in the treaty and country-specific port laws ensure this illicit trade in a different format (Shittu et al. 2021). One of the major reasons towards this is the lack of proper science-policy framework that ensures urban mining of e-waste. Even though there is a Waste Electrical and Electronic Equipment (WEEE) directive in the EU, country-specific rules/laws on e-waste, and legislation adhering to circular economy concepts, the topic of supply chain sustainability is rarely addressed.

Sustainability is now widely regarded as the next big revolution since the shift to digital is disrupting the way that business gets done. More businesses are viewing sustainability as a critical strategic issue (Burritt and Schaltegger 2014; Hassini et al. 2012). As a result, the body of knowledge on the topic has been expanding rapidly (Bansal and DesJardine, 2014). According to the sustainable supply chain management (SSCM) approach, concentrating on specific companies doesn't provide much insight into sustainability because supply chain-wide analysis is necessary to identify sustainability challenges (Silvestre 2016). This is due to the fact that certain focal corporations conceal unsustainable activities in remote areas of the supply chain to prevent negative publicity and reputational damage (Chan and Pun, 2010). As supply chains become increasingly complex and difficult to manage, researchers have been searching for useful approaches to deal with this complexity.

In the academic world, the scenario is different. Researchers are looking for practical methods to deal with the complexity of supply chains, which is becoming more and more challenging to manage. Several research groups are working in the area of e-waste supply chain sustainability. Jain et al. (2022) used the SCOR model to evaluate the sustainability performance of an e-waste closed loop SCN. They validated their model in the e-waste SCN of New Zealand. Wibowo et al. (2021) reported strategies for improving the sustainability of the e-waste SCN in Indonesia using MCDM techniques. They found that increasing public awareness and socialization of regulations are the two most important strategies required for increasing e-waste supply chain sustainability. Isernia et al. (2019) have presented the Italian scenario of e-waste reverse SCN with a focus on circular economy, which is interesting and timely. Debnath et al. (2020) performed the supply chain mapping of the e-waste sector and reported a generic global e-waste SCN with details. The SCN covers all stakeholders, including the informal sector, semi-informal sector, formal recycler, and second-hand market.

In summary, the current literature on the e-waste supply chain covers adequately the technical and trans-boundary issues along with the issues that are related to a specific country. However, sustainability of the e-waste business is not clear. Additionally, a systematic analysis of the e-waste SCN in order to achieve long-term sustainability is scant. Therefore, this study addresses these issues by studying the sustainability of the e-waste supply chain by thoroughly analyzing the e-waste supply chain characteristics, identifying issues and challenges, and suggesting improvement measures for sustainability.

6.2 METHODOLOGY

The chapter is developed using knowledge accumulated from published research articles, reports, and book chapters. Diverse online repositories and databases were explored such as ScienceDirect, Springer, Taylor & Francis, CRC Press, Nature, and Wiley in search of published literature. The following keywords and/or combination of keywords were employed for the purpose: 'e-waste,' 'supply chain management,' 'supply chain network,' 'machine learning,' 'game theory,' 'multi-criteria analysis,' 'fuzzy mathematics,' etc. The gathered articles and other sources were examined thoroughly to strain out the relevant ones, and the cross-references were also explored wherever necessary. Our search was limited within the last 15 years of published work. An in-depth assessment of the articles was done to acquire knowledge on the e-waste SCN, its issues and challenges, along with application of new and upcoming technologies in this sector. Finally, qualitative sustainability assessment was done based on prior knowledge and published literature. The references cited in each relevant literature and in this paper are reviewed and referred properly.

6.3 E-WASTE SUPPLY CHAIN MAPPING

The supply chain of any organization is a very important element, and the business sustainability of that particular organization depends on the efficiency and sustainability of the SCN. The elements and stakeholders pertaining to the SCN are required to be effective and robust in order to have a better impact on the business. The SCN of e-waste is a technically complex routine that is even more interesting due to the veracity and versatility of the supply chain dynamics (Debnath 2019).

The characteristics change from different geographical regions of the world to different economies. The main difference between developed and developing economies with respect to the supply chain is the presence of the informal sector and the second-hand market. Though heterogeneity is a big problem in both the economies, technically the developed economies are far ahead from the developing economies.

Ghosh et al. (2016) reported the first work in terms of e-waste and also reported the various issues and the challenges that were faced by the industry. Their main focus was working around the BRICS nations; however, it was surprising how generic the issues were and were not limited to the requirements that were set by the BRICS. It was found that the main problems came from the compliance requirements of the Basel Convention along with the movement of e-waste along the boundaries. With e-waste, there is a lot of informal collection of the same, which creates a huge issue in terms of accountability, and along with crude processing only makes the overall process more complicated. In the works of Cruz-Sotelo et al. (2017), he developed a mapping of a supply chain in the country of Mexico and also developed and showcased a legal framework that was sustainable in nature as well. In comparison, Baidya et al. (2020) showcased various issues in developing the same in India and used Analytic Hierarchical Process (AHP) to identify the main causes behind it. Future perspectives in the field of e-waste supply chains were put forward in the review written by Doan et al. (2019). Debnath (2019) showcased the network of movement of e-waste in India and also established the materialistic flow path in detail from consumer to a crude recycling methodology. India has a culture of repairing devices or selling them as refurbished devices until the point it can no longer be used and it is disposed of. This also reduces the overall impact range of recycling options, and a large number of literature options can be seen on this topic.

The major issues pertaining in the SCN of e-waste are the informal sector (Baidya et al. 2019; Hazra et al. 2011), trans-boundary shipment (Ghosh et al. 2014), formal waste collection (Ghosh et al. 2014), waste segregation (Baidya et al. 2019), and that for the developed countries there are loopholes of EPR (Debnath et al. 2015) and shipping of e-waste. The challenges related to e-waste SCN are (a) proper implementation of legislation (Debnath et al. 2015), (b) revision of rules (Debnath et al. 2014), (c) stopping e-waste smuggling, (d) formalization of the informal sector (Van Rooyen and Antonites 2007; Wang et al. 2020), (e) choosing the best available technology (Baidya et al. 2015), (f) boosting formal recycling (Debnath et al. 2014; UNEP 2013), and (g) introduction of advanced recycling fees (ARF) (Debnath et al. 2014). Though most of these are quite prominent in developing nations, developed nations also struggle to balance their supply chain dynamics.

6.4 APPLICATION OF NEW AND EMERGING TECHNOLOGIES FOR SUPPLY CHAIN ANALYSIS

6.4.1 Machine Learning

The digital industry has prospered and become an integral part of human society, as mentioned by Anh Khoa et al. (2020). The issue is that the industry, due to the heavy competition in it, has realized the need for innovation (Hussain et al. 2020). Therefore, when a device is becoming obsolete and is being replaced by a newer version, the overall time that is required in order to actually meet the demand increases because there is a dearth of raw materials that is being acquired by excessive mining, as stated in Mele et al. (2021). The same can be acquired by the means of recycling the raw material from old devices, as shown by Thorat and Shardul (2021). By learning the correlation between the various features in each of these scenarios, the machine can set up a prediction system that can predict the procedure for the best output. Machine learning can help in actually taking the composition of an input device and dividing it into the various elements that can be extracted from it. It will be a classification problem that can be handled by utilizing the LSA-MCA model, as mentioned by Chen et al. (2021). Machine learning algorithms can take the device type as an input with any parts missing and can estimate the various compositions that can be achieved at what percentage. Kannangara et al. (2018) mentioned and showcased a use of random forests and decision tree-based classifiers that can be used in order to understand the direction and system that takes into account demographic and socio-economic variables for pre-planning developments in this area.

6.4.2 Artificial Intelligence

There are a huge number of machine-learning algorithms, and depending upon which input that is provided, the proper algorithm needs to be chosen. Convolutional neural networks hence showcase a better understanding of making that choice by creating a densely connected network that can make the required classification. Even though machine learning models are pretty apt in getting statistical benefit from the data collected, the true advantage of using classification systems can be obtained by using them in conjunction with neural networks (Ahmed et al. 2020). Agarwal et al. (2020) mention the primary issues of handling e-waste. In developed countries there is an infrastructure present that can actually recycle the materials and gain an advantage; however, in the developing countries, due to lack of awareness, a lot of the reusable materials are wasted or dumped into landfills or waterbodies. This leads to toxification of these natural resources and hampers the life of both humans and other animals. Most countries also lack proper legal guidelines, and this leads to further complications. Batoo et al. (2021) hence offered the option of a behaviour-based swarm model that will take the data into consideration and try to use that to control and manage a heterogeneous system of vehicles for e-waste collection. They will not only be helpful in collecting the e-waste that is generated in densely populated areas of India and China, but also be effective in largely developed countries as well. Szwarc et al. (2021) showcased the nondeterministic polynomial hard nature of this problem. Since for most people disposing of such devices is a bigger issue, the lack of a system leads to people being exposed to wrongly dumping the device, which escalates the problem. They have proposed the use of a memetic approach in this scenario, which will lead to better optimized use and management of the resources. There will be a huge maintenance cost that will be there in having a fleet dedicated to this purpose, which also boosts up the overall importance. The convolutional neural network will be responsible for looking at each scenario and be responsible for choosing which process should be used in order to handle e-waste of a particular type based on its current situation.

6.4.3 Game Theory

Game theory is a type of mathematical model that aims to help in decision making by providing rewards to decisions that benefit the system and also penalizing wrong decisions. Game theory has gained a lot of popularity as it can assist in complicated decision making and works as an optimization model. Game theory tries to maximize the actions in the model that reap rewards and minimize the ones that are penalized. Game theory is used as an incentive to get more corporations interested in e-waste management, as shown in Li et al. (2020). China is one of the world's largest consumers of electronics. According to a survey conducted in 2018, China was responsible for around 5.73 million tonnes of e-waste. As suggested in the works of Zhao and Bai (2021), they showcased that by using game theory it may be possible to include the private corporations in China in order to actually invest in this domain. They realized that most people in the real world are not rational and respond better to a reward-punishment system. China set up a fund in 2012 to be spent on tackling the e-waste problem. China offered private corporations benefits from both the funds as well as tax deductions for maintaining proper practises. The manufacturers also had to go through a trial-and-error method in order to showcase the best options in terms of choosing to manage the waste, which showed a positive impact on the financials. Hou et al. (2020) used game theory in order to decide which model should be chosen: One in which a middle man collects all the second-hand air conditioners and either resells them back or breaks them down for parts, or established corporations to which consumers can give directly or the middle man can give directly in exchange for monetary gain. People prefer having a middle person to whom they can give and forget. Xuehua (2021) showcases another issue in recycling that comes with the low quality of the procedure. The overall final product does not have enough quality in order to make up for the missing elements. This is why they use decision trees in order to try and understand which methodology is the one to be chosen and come to the conclusion that both quasi and mixed forms of separation lead to degradation of quality, and only single file separation must be used.

6.4.4 Fuzzy Mathematics

In mathematics, a lot of the models are based on hypothetical scenarios that do not account for real-life situations. These are usually added later as a part of the error or bias that cannot be controlled. A simple example can be seen in the support vector machines. A hyperplane that perfectly classifies the majority of classes perfectly is not possible practically. However, fuzzy mathematics solves that by incorporating these factors into the model itself. The fuzzy logic, when applied to SVM, allows points to lie on the hyperplane itself, with each of the points incurring a cost. The problem therefore adds another layer of modelling the hyperplane such that it optimizes the cost, which practically makes it more accurate to real-life scenarios, as shown in Hassani et al. (2017). After a device has been collected, the next obvious step is to understand how to recycle it. The works of Yeh and Xu (2013) thereby showcase a sustainable approach to the planning of the recycled material by using fuzzy multicriteria decision making (MCDM). The sustainable factors that take into consideration look at the cost to implement a route as well as the carbon footprint that will be added as well as the recyclable product and the elements achieved from the same in order to understand whether it is a feasible option or not. The fuzzy MCDM option uses a multi-weighted approach by providing each of the various factors a weight based on a priority level and then uses that to measure the overall importance of a job or a route. This makes it a very valuable asset for e-recycling corporations in order to optimize their functionalities. Shumon et al. (2016) showcased this to be a reverse supply chain problem. This is collecting from the consumers and then getting it back to the corporation, and since there are multiple criteria here that need to be considered, the usage of a fuzzy AHP makes sense. If the corporation is implementing multiple systems to manage the e-waste, then this can be used to make a choice on the optimal one. Another approach has been suggested by Rani et al. (2020), who showcased the use of fuzzy logic in inventory management. This will allow them to develop and resell refurbished products, which will feed into a crowd that aims to actually not spend that much on recent upgrades. This can revise per-unit cost of the electronics as a refurbished product will only need a quarter of the parts compared to a brand new one. This reduces cost on the company's behalf and also reduces overall drain on natural resources. With the introduction of swarm-based behaviour models in order to collect e-waste, the importance of fuzzy logic in order to show the importance of the same has also improved, as shown by Batoo et al. (2021).

6.4.5 MCDM Techniques

MCDM techniques are very important and widely used tools for supply chain management. There are several instances where MCDM techniques have been used for supply chain management in the case of the e-waste sector. Material flow analysis is a major method that is used in order to understand the flow of the overall material from a usable product to a point where it can be recycled or to understand where it mostly ends up in terms of usability, as shown in Streicher-Porte et al. (2005). Even though this study helps to differentiate between a flow path of a recyclable resource and the part which is not, termed e-waste, there is a lack of understanding of a generalized supply chain structure. An AHP MCDM was put forward by Sharma and Pratap (2013) with a sole aim to optimize the supply chain delivery network, which will look at both the qualitative and quantitative factors. In the application of a system that can make decisions based on the various issues in all political, technical, environmental, and economical factors, Ciocoiu et al. (2011) formulated and put forward an analytic hierarchy process. If a Quality Function Deployment (QFD) with a two-stage deployment system can be constructed, it will help in the strategic adjustment of the model in an e-waste management system, as shown by Yang et al. (2011). The usage of the QFD was made in the works of

Ghosh et al. (2014), who highlighted the various issues in the e-waste supply chain. The main issues in the e-waste supply chain were narrowed down to the collection of the waste, the availability of technology, and storage of the materials as the primary concerns. Prakash and Barua (2015) employed a combination of fuzzy-AHP and TOPSIS for prioritization of solutions for reverse logistics in a stochastic environment. Bal and Satoglu (2018) used goal programming to model the reverse logistics network of e-waste for different scenarios following the triple-bottom line approach of sustainability. Kumar and Dixit (2018) combined interpretive structural modelling (ISM) and Decision-Making Trail and Evaluation Laboratory (DEMATEL) to identify different constraints in the e-waste supply chain in the Indian scenario. Fuzzy-TOPSIS method was used to rank and prioritize pathways to e-waste formalization systems in Ghana. The study also identified that economic and financial constraints are the most important barricades towards e-waste formalization in Ghana (Chen et al. 2020). Baidya et al. (2020) have used a combined AHP-QFD method to prioritize different issues and challenges prevailing along the e-waste supply chain. They validated their findings through two case studies from two developing countries – India and China. The study compared the supply chain network of India and China and discussed the sustainability aspects in a qualitative manner. Recently, a DEMATEL coupled with Grey Concept was used to identify the causal relationship of different strategies of e-waste management. Hence, it can be seen that there is an inclination of the researchers towards using AHP, TOPSIS, coupled with other techniques such as fuzzy compared to others such as VIKOR, PROMETHEE, DEMATEL, etc. These have been slowly growing in the past couple of years or so.

6.4.6 Blockchain

Bag et al. (2020) showcased that one of the issues in handling of e-waste is logistics management. This is where blockchain can come into play by creating a ledger that has high transparency and is resistive to change. They also outline 15 issues that the current market faces in the implementation of blockchain. The first is a lack of a proper collection system. Obsolete devices are usually given as hand-me-downs and also in most cases just discarded. In cases where they are being recycled, it can be difficult to have a tracking system in place because of low quality of the device, lack of proper infrastructure, and generally low yielding returns in terms of cost invested to implement such a system. Xu et al. (2019) also proposed a different direction. The massive usage of devices meant that large-scale production was happening, which saw a very steep drop in quality. However, in case a quality issue was found, it was difficult to find the person liable, which is why he proposed a blockchain-based framework that will provide certificates at various levels over the responsibility handled in the parts of the device that will lead to improved quality and longer durability and reduced waste generation. Poongodi et al. (2020) also showcased that even though lack of infrastructure and tracking of discarded items can be difficult in general, in the newly developed smart cities it can be done. With cheap microprocessors being implemented into the trash cans and use of QR and NFC codes, identities of devices can be ingrained into the structure of the devices to further have stronger identities that can be used as a tracking system in a logistic sense. Lamichhane (2017) showed that the blockchain system is very resistant to changes once an entry has been added to this. This makes it so that as the product is moving through the system, starting from collection to the final sorting centre, the product can be tracked more accurately. Blocks that are made on the linked list data structure and utilize hashing and distribution in order to store the network information and at the same time hold both transparency and strong understanding of tracking the product. This is important in this industry as the number of items and their source plays a major role.

6.5 SUSTAINABILITY ASPECTS OF E-WASTE SCN

6.5.1 Computing Techniques for E-waste SCN Sustainability

One of the primary issues of e-waste is the sheer scale of it. The amount of menial labour that will be required in the actions is large. This will include cleaning and indexing all the various parts that are usable and the parts that are not, shown by Ozdemir et al. (2021). Therefore, this gap can be filled in by making use of machine learning, which can handle the classification problems (Nowakowski and Pamula 2020). By combining the technologies of IoT and machine learning, the processes can be completed (Anh Khoa et al. 2020). However, there are still the concerns of how this scale can be implemented and the idea of actually getting the manufacturers on board. This is where game theory can be implemented in order to add incentives to make this process lucrative Gu et al. (2005). Game theory can be used in order to find motivational factors that will make the difference (Zhang et al. 2020). In implementation of a said game theory model, it is important to make use of fuzzy logic as real life is not going to be as straight as a yes/no decision, and fuzzy logic will make the difference (Chen et al. 2020). Similarly, MCDM will be looking at the multiple factors in the decision-making chain to make a choice, as shown by Riahee and Zandieh (2018). One of the issues in the e-supply chain is accountability as well as keeping a transactional record that is easy to access on a large scale. That is where blockchain comes in. Chen et al. (2021) showed that blockchain will be able to go ahead and store this information on a peer-to-peer network, and due the nature of it, resisting change will be an important factor. Having a system that can track the movement of a recyclable e-waste over the complicated system will be a complicated endeavour that needs to be undertaken, as shown by Gopalakrishnan et al. (2021). It will create a secure and well-designed system, which will help in the overall methodology of tracking the movement and what the end product is in its lifecycle. In the e-waste supply chain, one of main challenges is working with it, to be able to separate out the portion that can be reused. However, this setup can have a very positive impact and reduce the overall mining demands of the industries. The high mining requirements lead the earth being mined constantly for resources, which takes years to build back up, leaving lower and lower resources for the future generations.

6.5.2 Sustainability Analysis

There are many factors governing sustainability of e-waste SCN, e.g., supply uncertainty, waste quality (heterogeneity), logistics support, intervention from a number of stakeholders, untrained manpower, appropriate recycling technology at an affordable cost, value addition process, trans-boundary shipment, compliance to local environmental and other legislative requirements, demand of recycled products and equipment for reuse, disposal of residues to landfills, and landfill costs. Hence, it is complex and will require a lot of trustworthy data and subsequent analysis for tracing the true sustainability. This is beyond the scope of this work; hence, we restrict the discussion within the qualitative territory. As stated, in the Brundtland report, the three pillars of sustainability are considered for subsequent discussion (Figure 6.2). The following discussion follows the qualitative sustainability analysis process present in literature (Ghosh et al. 2018; Debnath et al. 2021).

6.5.2.1 Environmental Sustainability

Any procedure that provides the least possible environmental damage is said to be environmentally sustainable. As mentioned before, due to lack to reliable data, Life Cycle Analysis (LCA) is beyond the scope of the work. So, using LCA's guiding principles, we present a qualitative study. We also take e-waste supply chain uncertainty into consideration. The following are revealed:

i. Using soft computation techniques, MCDM techniques and advancing techniques such as machine learning, it is expected to help the smooth management of e-waste, especially the post collection stage. Smart algorithms and programs can be employed to optimize processes and transportation issues, which can help in reducing the carbon footprint (Yang et al. 2020). As a result, it will affect the following categories: global warming potential, eutrophication, human toxicity, and natural resource depletion.
ii. Efficient management of e-waste will also ensure the channelization of e-waste towards formal recycling units, which in turn is expected to reduce the transboundary movement issue of e-waste. This will ensure environmental sustainability from the demand aspect. In addition, this will reduce environmental load on shipping and subsequent rudimentary recycling. The possible affected categories are climate change, acidification, photochemical smog formation, and particulate matter formation.
iii. Implementation of technologies such as blockchain will enhance the monitoring of the supply chain, which may be particularly useful in developing nations to reduce the channelization of e-waste to the informal sector. This will reduce the environmental impact as the following categories are possible to be affected – global warming potential, acidification, natural resource depletion, photochemical smog formation, human health hazards, and urban land occupation. However, the conundrum remains, that too much application of blockchain may lead to higher e-waste generation (De Vries and Stoll 2021). Hence, this aspect needs to be evaluated separately (Rieger et al. 2022).

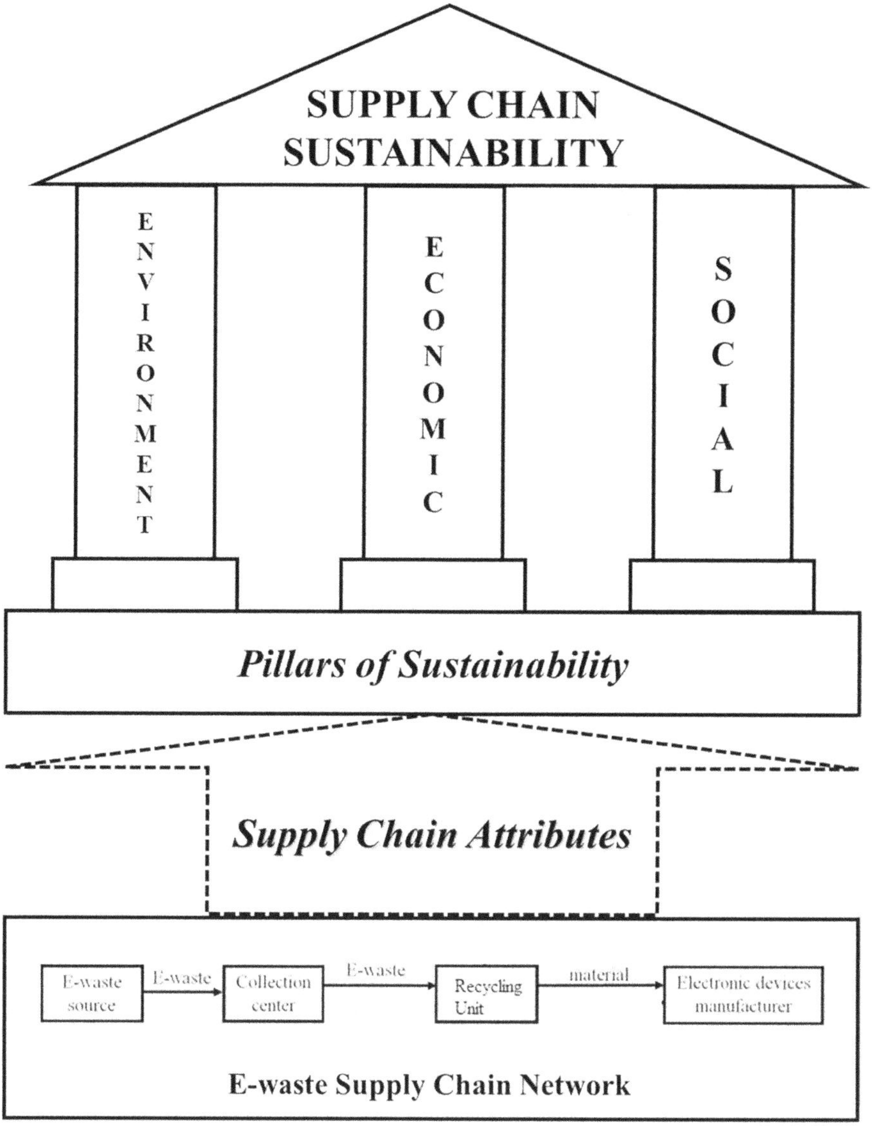

FIGURE 6.2 Supply chain sustainability.

6.5.2.2 Economic Sustainability

Economic sustainability is the most important aspect of supply chain sustainability. The dynamics of e-waste SCN is highly affected by price fluctuations and market volatility. As mentioned by Anand and Sen (2000), economic sustainability is not always straightforward as it seems. As a result, determination of economic sustainability could be highly complex and mathematically exhaustive.

To keep it simple, we present a generalized analysis following the systems approach. The analysis reveals the following points:

i. Implementation of the discussed techniques will reduce transportation issues and increase the ease of operations management. This will reduce a great deal of cost.
ii. A sustainable e-waste SCN will ensure resource circulation, which will in turn be helpful in boosting the economy.
iii. E-waste SCN is often affected by the market volatility and price fluctuations. This is often a major concern towards business sustainability. Attending to these issues through the discussed technologies will be able to handle these uncertainties wisely, thereby ensuring economic stability.

6.5.2.3 Social Sustainability

The Triple Bottom Line (TBL), which asserts that all three performance dimensions – economic, social, and environmental – are interconnected and of equal importance, is at the heart of the sustainability and SSCM approaches (Elkington 2002). However, the social component is frequently ignored in SSCM investigations (Abbasi, 2017; Ashby et al. 2012; Wu and Pagell, 2011), undervalued, underexplored, and under-theorized (Pullman et al. 2009; Silvestre 2016). Perhaps this neglect can be justified by the fact that businesses find it more difficult to address social issues than environmental ones (Ashby et al. 2012), or by the fact that sustainability, according to the broad TBL definition, is a theoretical concept with little application in the real world (Brandenburg et al. 2014).

In reality, social sustainability is tricky and variable based on location, aspects, industry, religion, climate, and many more. Hence, we present the possible scenarios of social development from a global aspect. The outcomes are provided below.

i. Social sustainability is an important aspect of supply chain sustainability. A lot of it depends on the internal operations as well as on the demand side. For example, workers are the backbone of the internal operations. Worker strike, wage issues, etc., can hamper the overall workflow; hence, it is important to nurture these micro-management aspects to maintain overall equilibrium.
ii. One important lesson from the COVID-19 pandemic is that supply chain sustainability can be easily disrupted due to unforeseen events. Creating resilient SCN for e-waste is hence important to tackle such uncertainties. More research on this direction is anticipated.
iii. Recent studies on SME supply chain sustainability have identified threats evolving from the social angle, which needs to be taken seriously; otherwise, the dynamics of the SCN might lead to bankruptcy (Debnath et al., 2022). The same can happen for e-waste SCN as it is much more volatile compared to SME SCN.

6.6 CONCLUSION

Supply chain sustainability is an emerging and interesting area, especially for the waste management sector. In this study, an in-depth analysis of e-waste supply chain literature has been carried out with a focus on soft computing techniques, MCDM techniques, machine learning, and blockchain technology. The analysis revealed that the penetration of such techniques is beneficial for the e-waste management sector. Although the conundrum remains with blockchain technology as it may increase the generation of e-waste, the sustainability analysis provides the current aspects of e-waste SCN based on the three pillars of sustainability, i.e., environmental, economic, and social. The analysis also provides research directions for better understanding of the paradigm shift of e-waste supply chain sustainability. The findings will help researchers and supply chain managers in future works, including research development and decision making.

REFERENCES

Abbasi, Maisam. "Towards socially sustainable supply chains–themes and challenges." *European Business Review* (2017).

Anand, Sudhir, & Sen, Amartya (2000). Human Development and Economic Sustainability. World Development, 28, 2029–204910.1016/s0305-750x(00)00071-1.

Agarwal, Vernika, Shivam Goyal, and Sanskriti Goel. "Artificial intelligence in waste electronic and electrical equipment treatment: Opportunities and challenges." In *2020 International Conference on Intelligent Engineering and Management (ICIEM)*, pp. 526–529. IEEE, 2020.

Ahmed, Alim Al Ayub, and A. B. M. Asadullah. "Artificial intelligence and machine learning in waste management and recycling." *Engineering International* 8, no. 1 (2020): 43–52.

Anh Khoa, Tran, Cao Hoang Phuc, Pham Duc Lam, Le Mai Bao Nhu, Nguyen Minh Trong, Nguyen Thi Hoang Phuong, Nguyen Van Dung, Nguyen Tan-Y, Hoang Nam Nguyen, and Dang Ngoc Minh Duc. "Waste management system using iot-based machine learning in university." *Wireless Communications and Mobile Computing* 2020 (2020).

Ashby, Alison, Mike Leat, and Melanie Hudson-Smith. "Making connections: a review of supply chain management and sustainability literature." *Supply Chain Management: An International Journal* 17, no. 5 (2012): 497–516.

Bag, Surajit, Dmitriev Aleksandr Viktorovich, Atul Kumar Sahu, and Anoop Kumar Sahu. "Barriers to adoption of blockchain technology in green supply chain management." *Journal of Global Operations and Strategic Sourcing* (2020).

Baidya, Rahul, Debnath, Biswajit, Ghosh, Sadhan Kumar, & Rhee, Seung-Whee (2020). Supply chain analysis of e-waste processing plants in developing countries. *Waste Management & Research*, 38, 173–18310.1177/0734242x19886633.

Bal, Alperen, & Satoglu, Sule Itir (2018). A goal programming model for sustainable reverse logistics operations planning and an application. *Journal of Cleaner Production*, 201, 1081–109110.1016/j.jclepro.2018.08.104.

Baldé, C. P., Forti, V.,Gray, V., Kuehr, R., & Stegmann, P. *The global e-waste monitor 2017: Quantities, flows and resources*. United Nations University, International Telecommunication Union, and International Solid Waste Association, 2017.

Bansal, Pratima, and Mark R. DesJardine. "Business sustainability: It is about time." *Strategic Organization* 12, no. 1 (2014): 70–78.

Batoo, Khalid Mujasam, Saravanan Pandiaraj, Muthumareeswaran Muthuramamoorthy, Emad H. Raslan, and Sujatha Krishnamoorthy. "Behavior-based swarm model using fuzzy controller for route planning and E-waste collection." *Environmental Science and Pollution Research* (2021): 1–15.

Brandenburg, Marcus, Kannan Govindan, Joseph Sarkis, and Stefan Seuring. "Quantitative models for sustainable supply chain management: Developments and directions." *European Journal of Operational Research* 233, no. 2 (2014): 299–312.

Burritt, Roger, and Stefan Schaltegger. "Accounting towards sustainability in production and supply chains." *The British Accounting Review* 46, no. 4 (2014): 327–343.

Chan, Jenny, and Ngai Pun. "Suicide as protest for the new generation of Chinese migrant workers: Foxconn, global capital, and the state." *The Asia-Pacific Journal* 37, no. 2 (2010): 1–50.

Chen, Dehu, Daniel Faibil, and Martin Agyemang. "Evaluating critical barriers and pathways to implementation of e-waste formalization management systems in Ghana: A hybrid BWM and fuzzy TOPSIS approach." *Environmental Science and Pollution Research* 27, no. 35 (2020): 44561–44584.

Chen, Jie, Shoujun Huang, S. BalaMurugan, and G. S. Tamizharasi. "Artificial intelligence based e-waste management for environmental planning." *Environmental Impact Assessment Review* 87 (2021): 106498.

Ciocoiu, C.N., Colesca, S. E., andBurcea, S. (2011, July). An AHP approach to evaluate the implementation of WEEE management systems. In *Recent Researches in Environment, Energy Planning and Pollution-Proc. of the 5th WSEAS Int. Conf. on Renewable Energy Sources, RES* (Vol. 11, pp. 233–238).

Cruz-Sotelo, Samantha, Ojeda-Benítez, Sara, Jáuregui Sesma, Jorge, Velázquez-Victorica, Karla, Santillán-Soto, Néstor, García-Cueto, O., Alcántara Concepción, Víctor, & Alcántara, Camilo (2017). E-Waste Supply Chain in Mexico: Challenges and Opportunities for Sustainable Management. *Sustainability*, 9, 503 10.3390/su9040503.

De Vries, Alex, and Christian Stoll. "Bitcoin's growing e-waste problem." *Resources, Conservation and Recycling* 175 (2021): 105901.

Debnath, Biswajit. Towards Sustainable E-Waste Management Through Industrial Symbiosis: A Supply Chain Perspective, *Industrial Symbiosis for the Circular Economy: Operational Experiences, Best Practices and Obstacles to a Collaborative Business Approach*, pp. 87–102, 2020.

Debnath, B. Sustainability of WEEE recycling in India. *Re-Use and Recycling of Materials*, pp. 15–32. River Publishers, 2022.

Debnath, Biswajit, Adrija Das, Ankita Das, Rohit Roy Chowdhury, Saswati Gharami, and Abhijit Das. "Edge Computing-Based Smart Healthcare System for Home Monitoring of Quarantine Patients: Security Threat and Sustainability Aspects." In *Intelligent Modeling, Prediction, and Diagnosis from Epidemiological Data*, pp. 189–210. Chapman and Hall/CRC, 2021.

Debnath, B., Baidya, R., Biswas, N. T.,Kundu, R., and Ghosh, S. K."E-waste Recycling as Criteria for Green Computing Approach: Analysis by QFD Tool." In *Computational Advancement in Communication Circuits and Systems: Proceedings of ICCACCS 2014*, pp. 139–144. Springer India, 2015.

Debnath, B., Baidya, R., &Ghosh, S.K. Simultaneous analysis of WEEE management system focusing on the supply chain in India, UK and Switzerland. *International Journal of Manufacturing & Industrial Engineering*, 2, no. 1 (2015): 16–20.

Debnath, Biswajit, El-Hassani, Rihab, Chattopadhyay, Amit K., Kumar, T. Krishna, Ghosh, Sadhan K., & Baidya, Rahul (2022). Time evolution of a Supply Chain Network: Kinetic modeling. *Physica A: Statistical Mechanics and its Applications*, 607, 128085. 10.1016/j.physa.2022.128085

Dias, Pablo, Md Tasbirul Islam, Bin Lu, Nazmul Huda, and Andréa M. Bernarde. "e-waste transboundary movement regulations in various jurisdictions." *Electronic Waste: Recycling and Reprocessing for a Sustainable Future* (2022a): 33–59.

Dias, Pablo, Andréa M. Bernardes, and Nazmul Huda. "e-waste management and practices in developed and developing countries." *Electronic Waste: Recycling and Reprocessing for a Sustainable Future* (2022b): 15–32.

Doan, Linh Thi Truc, Yousef Amer, Sang-Heon Lee, Phan Nguyen Ky Phuc, and Luu Quoc Dat. "E-waste reverse supply chain: A review and future perspectives." *Applied Sciences* 9, no. 23 (2019): 5195.

Dwivedy, Maheshwar, & Mittal, R.K. (2010). Estimation of future outflows of e-waste in India. Waste Management, 30, 483–49110.1016/j.wasman.2009.09.024.

Elkington, John. *Cannibals with Forks: The Triple Bottom Line of 21st Century Business.* Capstone, Oxford, 2002.

Forti, Vanessa, Cornelis P. Balde, Ruediger Kuehr, and Garam Bel. "The global e-waste monitor 2020: Quantities, flows and the circular economy potential." (2020).

Ghosh, S. K., Baidya, R., Debnath, B.,Biswas, N. T., & De D, L. M. E-waste supply chain issues and challenges in India using QFD as analytical tool. In *Proceedings of international conference on computing, communication and manufacturing, ICCCM* (pp. 22–23), 2014, November.

Ghosh, Sadhan Kumar, Debnath, Biswajit, Baidya, Rahul, De, Debashree, Li, Jinhui, Ghosh, Sannidhya Kumar, Zheng, Lixia, Awasthi, Abhishek Kumar, Liubarskaia, Maria A., Ogola, Jason S., & Tavares, André Neiva (2016). Waste electrical and electronic equipment management and Basel Convention compliance in Brazil, Russia, India, China and South Africa (BRICS) nations. *Waste Management & Research: The Journal for a Sustainable Circular Economy*, 34, 693–70710.1177/0734242x16652956.

Ghosh, Anaya, Biswajit Debnath, Sadhan Kumar Ghosh, Bimal Das, and Jyoti Prakas Sarkar. "Sustainability analysis of organic fraction of municipal solid waste conversion techniques for efficient resource recovery in India through case studies." *Journal of Material Cycles and Waste Management* 20, no. 4 (2018): 1969–1985.

Gopalakrishnan, Praveen Kumare, John Hall, and Sara Behdad. "Cost analysis and optimization of Blockchain-based solid waste management traceability system." *Waste Management* 120 (2021): 594–607.

Gu, Q. L., T. G. Gao, and L-Sh Shi. "Price decision analysis for reverse supply chain based on game theory." *Systems Engineering-Theory & Practice* 25, no. 3 (2005): 20–25.

Hassani, Hossein, Jafar Zarei, Mohammad Mehdi Arefi, and Roozbeh Razavi-Far. "zSlices-based general type-2 fuzzy fusion of support vector machines with application to bearing fault detection." *IEEE Transactions on Industrial Electronics* 64, no. 9 (2017): 7210–7217.

Hassini, Elkafi, Chirag Surti, and Cory Searcy. "A literature review and a case study of sustainable supply chains with a focus on metrics." *International Journal of Production Economics* 140, no. 1 (2012): 69–82.

Hazra, J., Sarkar, A., & Sharma, V. E-waste supply chain management in India: Opportunities and challenges. *Clean India Journal*, 7 (2011).

Hou, Jiayue, Qun Zhang, Shanying Hu, and Dingjiang Chen. "Evaluation of a new extended producer responsibility mode for WEEE based on a supply chain scheme." *Science of The Total Environment* 726 (2020): 138531.

Hussain, Ayaz, Umar Draz, Tariq Ali, Saman Tariq, Muhammad Irfan, Adam Glowacz, Jose Alfonso Antonino Daviu, Sana Yasin, and Saifur Rahman. "Waste management and prediction of air pollutants using IoT and machine learning approach." *Energies* 13, no. 15 (2020): 3930.

Isernia, Raffaele, Renato Passaro, Ivana Quinto, and Antonio Thomas. "The reverse supply chain of the e-waste management processes in a circular economy framework: Evidence from Italy." *Sustainability* 11, no. 8 (2019): 2430.

Jain, Vipul, Sameer Kumar, Amirhossein Mostofi, and Mojtaba Arab Momeni. "Sustainability performance evaluation of the E-waste closed-loop supply chain with the SCOR model." *Waste Management* 147 (2022): 36–47.

Ji, Xuehua. "The game model of signal transmission between e-waste recycling platform and recycler." *Journal of Physics: Conference Series*, 1774, no. 1(2021): 012062. IOP Publishing.

Kannangara, Miyuru, Rahul Dua, Leila Ahmadi, and Farid Bensebaa. "Modeling and prediction of regional municipal solid waste generation and diversion in Canada using machine learning approaches." *Waste Management* 74 (2018): 3–15.

Kumar, Ashwani, & Dixit, Gaurav (2018). An analysis of barriers affecting the implementation of e-waste management practices in India: A novel ISM-DEMATEL approach. *Sustainable Production and Consumption*, 14, 36–5210.1016/j.spc.2018.01.002.

Lamichhane, Manish. "A smart waste management system using IoT and blockchain technology." (2017).

Li, Hui-Jia, Qian Wang, Shenfeng Liu, and Jun Hu. "Exploring the trust management mechanism in self-organizing complex network based on game theory." *Physica A: Statistical Mechanics and Its Applications* 542 (2020): 123514.

Mele, Marco, and Cosimo Magazzino. "Pollution, economic growth, and COVID-19 deaths in India: A machine learning evidence." *Environmental Science and Pollution Research* 28, no. 3 (2021): 2669–2677.

Nowakowski, Piotr, and Teresa Pamuła. "Application of deep learning object classifier to improve e-waste collection planning." *Waste Management* 109 (2020): 1–9.

Ozdemir, Merve Erkinay, Zaara Ali, Balakrishnan Subeshan, and Eylem Asmatulu. "Applying machine learning approach in recycling." *Journal of Material Cycles and Waste Management* (2021): 1–17.

Prakash, Chandra, & Barua, M.K. (2015). Integration of AHP-TOPSIS method for prioritizing the solutions of reverse logistics adoption to overcome its barriers under fuzzy environment. *Journal of Manufacturing Systems*, 37, 599–61510.1016/j.jmsy.2015.03.001.

Poongodi, M., Mounir Hamdi, V. Vijayakumar, Bharat S. Rawal, and Ma Maode. "An effective electronic waste management solution based on blockchain smart contract in 5G Communities." In *2020 IEEE 3rd 5G World Forum (5GWF)*, pp. 1–6. IEEE, 2020.

PULLMAN, MADELEINE E., MALONI, MICHAEL J., & CARTER, CRAIG R. (2009). FOOD FOR THOUGHT: SOCIAL VERSUS ENVIRONMENTAL SUSTAINABILITY PRACTICES AND PERFORMANCE OUTCOMES. *Journal of Supply Chain Management*, 45, 38–5410.1111/j.1745-493x.2009.03175.x.

Rani, Smita, Rashid Ali, and Anchal Agarwal. "Fuzzy inventory model for new and refurbished deteriorating items with cannibalisation in green supply chain." *International Journal of Systems Science: Operations & Logistics* (2020): 1–17.

Riahee, Mina, and Mostafa Zandieh. "Disassembly line balancing based on the Kano model and Fuzzy MCDM methods, the case: E-waste recycling line." *Industrial Management Studies* 16, no. 49 (2018): 1–36.

Rieger, A., Roth, T., Sedlmeir, J., and Fridgen, G. (2022). We need a broader debate on the sustainability of blockchain. *Joule* 6(6), 1137–1141.

Sharma, S. and Pratap, R. A case study of risks optimization using Ahp method. *Journal of Scientific and Research Publications*, 3, no. 10 (2013): 1–16.

Shittu, Olanrewaju S., Ian D. Williams, and Peter J. Shaw. "Global E-waste management: Can WEEE make a difference? A review of e-waste trends, legislation, contemporary issues and future challenges." *Waste Management* 120 (2021): 549–563.

Shumon, Md Rezaul Hasan, Shamsuddin Ahmed, and Shameem Ahmed. "Fuzzy analytical hierarchy process extent analysis for selection of end of life electronic products collection system in a reverse supply chain." *Proceedings of the Institution of Mechanical Engineers, Part B: Journal of Engineering Manufacture* 230, no. 1 (2016): 157–168.

Silvestre, Bruno. "Sustainable supply chain management: Current debate and future directions." *Gestão & Produção* 23 (2016): 235–249.

Streicher-Porte, Martin, Widmer, Rolf, Jain, Amit, Bader, Hans-Peter, Scheidegger, Ruth, & Kytzia, Susanne (2005). Key drivers of the e-waste recycling system: Assessing and modelling e-waste processing in the informal sector in Delhi. *Environmental Impact Assessment Review*, 25, 472–49110.1016/j.eiar.2005.04.004.

Szwarc, Krzysztof, Piotr Nowakowski, and Urszula Boryczka. "An evolutionary approach to the vehicle route planning in e-waste mobile collection on demand." *Soft Computing* 25, no. 8 (2021): 6665–6680.

Thorat, Shardul. "Review of life-cycle analysis of e-waste in India." *International Journal of Modern Agriculture* 10, no. 2 (2021): 838–857.

UNEP. Report. Available at:https://www.unodc.org/documents/ data-and-analysis/Studies/TOCTA_EAP_web.pdf (accessed 12 October 2015), 2013.

Van Rooyen, E. J. and Antonites, A. J.. "Formalising the informal sector: a case study on the City of Johannesburg." *Journal of Public Administration* 42, no. 3 (2007): 324–346.

Wang, Juntao, Li, Wenhua, Mishima, Nozomu, & Adachi, Tsuyoshi (2020). Formalisation of informal collectors under a dual-recycling channel: A game theoretic approach. *Waste Management & Research*, 38, 576–58710.1177/0734242x19897125.

Wibowo, Nurhadi, Jerry Kuswara Piton, Rahmat Nurcahyo, Djoko Sihono Gabriel, Farizal Farizal, and Alfian Ferdiansyah Madsuha. "Strategies for improving the e-waste management supply chain sustainability in Indonesia (Jakarta)." *Sustainability* 13, no. 24 (2021): 13955.

Wu, Zhaohui, and Mark Pagell. "Balancing priorities: Decision-making in sustainable supply chain management." *Journal of Operations Management* 29, no. 6 (2011): 577–590.

Xu, Xiaolin, Fahim Rahman, Bicky Shakya, Apostol Vassilev, Domenic Forte, and Mark Tehranipoor. "Electronics supply chain integrity enabled by blockchain." *ACM Transactions on Design Automation of Electronic Systems (TODAES)* 24, no. 3 (2019): 1–25.

Yang, Chang-Lin, Huang, Rong-Hwa, & Ke, Wen-Chuan (2011). Applying QFD to build green manufacturing system. *Production Planning & Control*, 23, 145–15910.1080/09537287.2011.591632.

Yang, Haichao, Sheng Zhang, Weifeng Ye, Yuzhe Qin, Meng Xu, and Lidong Han. "Emission reduction benefits and efficiency of e-waste recycling in China." *Waste Management* 102 (2020): 541–549.

Yeh, Chung-Hsing, and Yan Xu. "Sustainable planning of e-waste recycling activities using fuzzy multicriteria decision making." *Journal of Cleaner Production* 52 (2013): 194–204.

Zhang, Deyuan, Yushu Cao, Yingjie Wang, and Guoyu Ding. "Operational effectiveness of funding for waste electrical and electronic equipment disposal in China: An analysis based on game theory." *Resources, Conservation and Recycling* 152 (2020): 104514.

Zhao, Xiaomin, and Xueli Bai. "How to motivate the producers' green innovation in WEEE recycling in China?–An analysis based on evolutionary game theory." *Waste Management* 122 (2021): 26–35.

7 A Blockchain-Based Solution for Effective E-waste Management in Sustainable Smart Cities

A Supply Chain Perspective

Ankita Das, Adrija Das, Biswajit Debnath, Soumyajit Chatterjee, and Abhijit Das

7.1 INTRODUCTION

The biggest constraints facing the waste management sector are the volume of waste generated and the deficiency of a transparent management system. In 2016, 2.01 billion tonnes of waste were generated worldwide annually, which is assumed to rise to 3.40 billion tonnes by 2050 (Kaza et al. 2018). Electronic waste (e-waste) is growing at thrice the pace compared to other waste streams (Kaza et al. 2018). E-waste is any abandoned household or commercial item that contains circuitry or electrical components, as well as a power or battery supply. E-waste includes televisions, air conditioners, refrigerators, fluorescent and mercury-containing bulbs, computers, printers, photocopiers, mobile phones, and earphones (Widmer et al. 2005; Wath et al. 2010). The amount of e-waste generated worldwide is increasing at an alarming rate. In 2016, all countries produced a total of 44.7 million metric tonnes (Mt). By 2021, this figure is expected to rise to 52.2 Mt (Balde et al. 2017). India, with one of the world's fastest-growing electronics industries, is one of the top contributors of e-waste in Asia. Waste electrical and electronic equipment (WEEEs) are made up of a variety of components, some of which contain toxic substances that, if not handled properly, can have a negative impact on human health and the environment (Wath et al. 2010; Annamalai 2015). These hazards are frequently caused by inefficient recycling and disposal processes. Lead, mercury, arsenic, cadmium, and flame retardants are among the hazardous elements found in e-waste (Brindhadevi et al. 2022). Ferrous and nonferrous metals, plastics, glass, printed circuit boards, rubber, and other materials are also included in e-waste (Debnath 2022). The presence of precious metals such as gold, silver, and platinum in this category of waste entices the informal sector to pursue unscientific practises (Terazono et al. 2017). Extraction methods such as burning, acid baths, etc., lead

 DOI: 10.1201/9781003317050-9

to improper disposal of untreated e-waste in landfills, which allows harmful elements to enter land and water resources (Annamalai 2015; Orlins & Guan 2016). Many developed countries dispose of their e-waste by illegally exporting it to developing countries such as India, which serve as hubs for improper recycling (Dias et al. 2022). The health of the uneducated and unskilled labourers involved in the segregation and dismantling of WEEEs is also jeopardized due to the poor practises used during this process (Parajuly et al. 2018).

E-waste management (EWM) is the process of properly disposing of e-waste. The first step is to collect e-waste from consumers, which is then sorted into reusable and non-reusable products. Reusable items are kept for resale, while non-reusable items are disassembled. Dismantled non-reusable parts are shredded and separated into multiple rounds. They are either recycled to be reused or safely disposed of after hazardous components have been properly treated (Debnath 2020). Recent studies on the application of blockchain in the waste management sector are increasing, including for e-waste. Very few studies have focused on repairability of e-waste and e-components. Under the current investigation, a blockchain-based solution for sustainable management of e-waste has been proposed with a focus on repairability of e-items and their components.

7.2 METHODOLOGY

7.2.1 Assumptions

This study focuses on developing a transparent and sustainable e-waste supply chain network. We consider e-waste generated from major waste generators. We also assume that the informal sector doesn't interfere here and hence neglected this case.

7.2.2 Chain Description

The considered chain is presented in Figure 7.1. At first, the e-waste generated is collected by collector-cum-dismantlers via third-party logistics. The repairable and refurbishable fraction is sent to the refurbisher. The refurbished products are transported to the second-hand market for sale. The unrepairable fraction is channelized to the formal recycler where it is processed in four major fractions – metal, glass, plastic, and reusable parts. The reusable part is routed to the refurbisher, while third-party recyclers recycle the respective fractions.

7.2.3 Blockchain Strategy

The blockchain is built following the strategy entailed below. Each segment is stored in a distinct block on the blockchain network. At the premises of collector-cum-dismantler the wastes are checked on the basis of repairability. Again, at the formal recycling facility, the waste is further checked based on each recyclable fraction. After processing the recyclable wastes are delivered to third-party recycling units. The coding was carried out using Python programming language version 3.0.

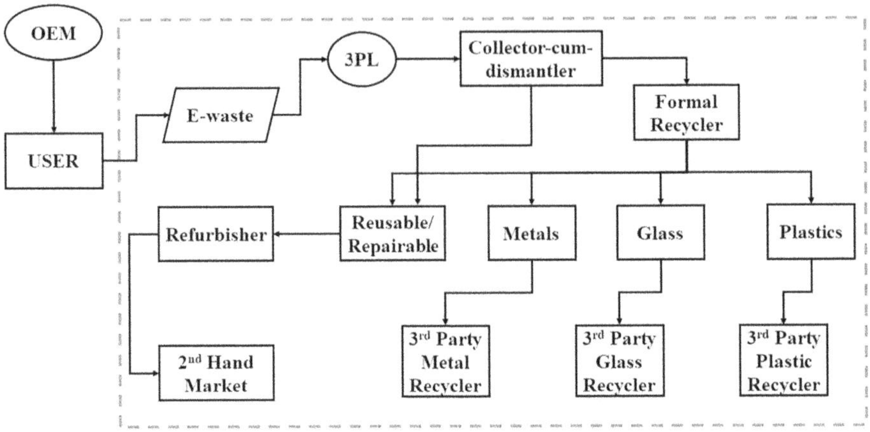

FIGURE 7.1 E-waste supply chain network considered for blockchain implementation.

7.3 RELEVANT LITERATURE ON E-WASTE AND BLOCKCHAIN

In these last two decades, numerous industries have been transformed by digital technologies (Andreoni et al. 2021). It began with the broadening of the Internet followed by the mobile technologies and ultimately smartphones. At present, artificial intelligence (AI), the Internet of Things (IoT), 3D printing, and blockchain technologies are conveying potentially disorderly solutions for sustainability and the environment (Rosário & Dias 2022).

Saberi et al. (2019) have scrutinized how blockchain technology and smart contracts can be supportive in approaching supply chain sustainability. Of late, it was observed from a meticulous review of the application of blockchain technology in supply chain management that the most fruitful researches are from the United States of America, China, and India. The study also offers future directions of blockchain research in supply chain management (Rejeb et al. 2021).

A novel smart EWM system was proposed by Sahoo and Halder (2020), by leveraging the power of blockchain technology and smart contract, considering both forward and reverse supply chains. The complete life cycle of e-products is recorded by the system beginning from their development as novel products to their dumping as e-waste, as well as their recycling back into raw materials. The researchers have extended their work by developing a prototype using solidity on the Ethereum platform. They established the fact that the use of blockchain technology enhanced accountability, lucidity, and conviction in the system (Sahoo et al. 2021). Chaudhury et al. (2021) have introduced an Ethereum-based smart contract blockchain architectural platform 'e-waste lens' for efficient EWM in India. The precise novelty of the blockchain architectural framework depends on the real-time tracking and scrutinizing of e-waste movement by the alarmed stakeholders in the EWM chain.

Akram et al. (2021) presented a reward system based on blockchain for solid waste management. In this system, LoRa-based sensors are tethered to waste-bins to observe the garbage level. These data are loaded to the cloud on runtime and consequently written on the blockchain. The individuals who dumped their waste in the bins are indistinctively rewarded based on the amount of waste.

Ahmad et al. (2021) displayed a blockchain-based solution for COVID-19 medical equipment waste management. This system utilizes Ethereum blockchain with Interplanetary File System (IPFS) for the administration of COVID-19 medical equipment data.

Recently, Khan and Ahmad (2022) proposed a blockchain-based IoT-enabled e-waste tracking and tracing system for smart cities that was tested using Ethereum. The system is aided by five smart contracts that verify and record the activities of users on the unalterable distributed ledger, which subsequently helps in the confirmation of the business processes executed by the participants, which are transparent, perceptible, and safe.

7.4 DISCUSSION AND ANALYSIS

7.4.1 Chain Description

Blockchain is a technology that allows for the sequential listing of transactions in a public digital ledger. Every area is seeing the emergence of blockchain, which can help to solve many global problems. If every electronic device has a digital identity that enables it to track and authenticate its origins, attributes, and ownership, a blockchain-driven system will provide clarity to e-waste flows and may undoubtedly resolve the issue of EWM in India. To provide value chain transparency and traceability, the e-waste supply chain network enabled by blockchain enables customers and collectors to disclose their methods of manufacturing. When attributes of e-waste are packaged as transactions and written on blockchains, the corresponding sub-consensus chain's mechanisms check and validate them to create a new block. A Merkle tree is used to hold the transaction's data, and its header includes information such as the hash of earlier blocks, an index, a timestamp, a nonce, and more.

7.4.2 Transaction and Block Structure

On the blockchain, all activities are documented as transactions. By merging transactions, it creates a candidate block from which the consensus layer can select.

7.4.2.1 Transaction Definition

Blockchain is a supplementary ledger. Data are copied and saved in real time on each node in the system since they are dispersed over several nodes. The price, asset, and ownership of a transaction are all recorded, confirmed, and instantly settled across all nodes when it is registered in the blockchain. A validated change that is registered on one ledger is simultaneously registered on all other copies of that ledger. Since every transaction is transparent and irrevocably recorded across all ledgers and made public, there is no requirement for independent third-party verification. The following tuples can be used to define the transaction in the suggested framework:

$$T_{X_{mn}} = \{Id_{mn}, H_{mn}, CT_{mn}, Ts_{mn}, I/O_{mn}\} \tag{6.1}$$

Here is a common transaction $T_{X_{mn}}$, which contains the hash value H_{mn}, enabling the transaction's encryption. The timestamp for the time lock of this particular transaction being received is Ts_{mn}, and the transaction index assigned by the blockchain is Id_{mn}. CT_{mn} is the definition of the current transaction. Input and output messages for transaction content are known as I/O_{mn}. The identifiers m and n in equation (6.1) stand for the transaction nodes and e-waste, respectively.

Any I/O_{mn} message must adhere to the following format:

```
I/Omn = {"Prev_out":[...]
         "out": [{"hash":" 6da9e1c2468d5 ... .."
                  "checkval": "a89d97e8a9a44 ...."
                  "get_tran":[...]}]
        }                                            (6.2)
```

A typical output consists of the characteristics *hash*, *checkval,* and *get_tran*, while an input typically contains data from the previous block in the network, marked by *Prev_out*. The functions *get_tran* and *checkval* encode data in hexadecimal and return a list of all network transactions, respectively.

7.4.2.2 Block Structure

Nodes broadcast their transactions over the network, and network peers will verify the signature and encrypt the transactions they got in one timeframe into a block for verification. Apart from the genesis or first block, the overall structure of the blocks in the network is shown in equation (6.3). Transparency of transactions is ensured via the flag characteristic that is present in each block.

$$Blck_m = \{Id_m, H_{prev}, Ts_m, Array[CT_m], \; nonce\} \tag{6.3}$$

Where

$$H_{prev} = Hash(Blck_{m-1})$$

There are no preceding hash values in the initial or genesis block. For the network's initial block, a random hexadecimal string is explicitly given as the hash value. The current transaction has been assigned to the genesis block as an arbitrary array. Sample code is given in Figure 7.2.

```
class Blk:
    def __init__(self, index,previous_hash,current_transaction,timestamp,nonce):
        self.index=index
        self.previous_hash=previous_hash
        self.current_transaction=current_transaction
        self.timestamp=timestamp
        self.nonce=nonce
        self.hash=self.hash_computation()

    def hash_computation(self):
        blk_string=json.dumps(self.__dict__,sort_keys=True)
        first_hash=sha256(blk_string.encode()).hexdigest()
        second_hash=sha256(first_hash.encode()).hexdigest()
        return second_hash
    def __str__(self):
        return str(self.__dict__)

class Blkchain:
    def __init__(self):
        self.chain=[]
        self.transactions=[]
        self.genblock()

    def __str__(self):
        return str(self.__dict__)
    def genblock(self):
        genblock=Blk('gen',0x0,[3,4,5,6],'datetime.now().timestamp()',0)
        genblock.hash=genblock.hash_computation()
        self.chain.append(genblock.hash)
        self.transactions.append(str(genblock.__dict__))

    def lastBlk(self):
        return self.chain[-1]
    def checkval(self,blk:Blk):
```

FIGURE 7.2 Sample code for blockchain.

7.4.3 Systematic Analysis

The three most significant methods are covered in this section. These are how blockchains are built. To secure each block in the chain, add new blocks to the chain, and retrieve all of the network's transactions: *Checkval*, *add_new_block*, and *get_Tran* are used, respectively. Output from the code is given in Figure 7.3.

7.4.4 Policy Aspects from the European Context

Waste management is one of the major concerns in Europe, including the notorious e-waste, often known as WEEE. The scenario of WEEE management has changed quite a lot in the last couple of years due to the pandemic, i.e., COVID-19, which tangled the supply chain scenario not only in Europe but around the world (Spieske et al. 2022). With an increase in work-from-home and online class scenarios, a rise in e-waste in the near future will not be a surprise. To tackle such issues, it is imperative to invest in strengthening the supply chain for sustainable operations management. One of the timely approaches taken by Europe is the adoption of

Customer:

{'index': 1, 'previous_hash': '82fc0b46f83fb54aef9484919f9edf74d16fdc90b165c16d7a667b79c1b00285', 'current_transaction': {'transactions': [{'timestamp': 1660389278.193925, 'Name': 'Smartphone', 'Repairability': 'yes', 'Size': 'small', 'from': 'Customer x', 'to': 'Transportation x', 'Digital signature': 'approaved', 'msg': 'This waste is in good order', 'FLAG': 'N'}, {'timestamp': 1660389278.193925, 'Name': 'Tab', 'Repairability': 'yes', 'Size': 'small', 'from': 'Customer x', 'to': 'Transportation x', 'Digital signature': 'approaved', 'msg': 'This waste is in good order', 'FLAG': 'N'}, {'timestamp': 1660389278.193925, 'Name': 'MixerGrinder', 'Repairability': 'No', 'Size': 'medium', 'from': 'Customer x', 'to': 'Transportation x', 'Digital signature': 'approaved', 'msg': 'This waste is in good order', 'FLAG': 'N'}, {'timestamp': 1660389278.193925, 'Name': 'Laptop', 'Repairability': 'yes', 'Size': 'medium', 'from': 'Customer x', 'to': 'Transportation x', 'Digital signature': 'approaved', 'msg': 'This waste is in good order', 'FLAG': 'N'}, {'timestamp': 1660389278.193925, 'Name': 'TV', 'Repairability': 'No', 'Size': 'large', 'from': 'Customer x', 'to': 'Transportation x', 'Digital signature': 'approaved', 'msg': 'This waste is in good order', 'FLAG': 'N'}, {'timestamp': 1660389278.193925, 'Name': 'Desktop-Computer', 'Repairability': 'No', 'Size': 'large', 'from': 'Customer x', 'to': 'Transportation x', 'Digital signature': 'approaved', 'msg': 'This waste is in good order', 'FLAG': 'N'}]}, 'timestamp': 'datetime.now().timestamp()', 'nonce': 17, 'hash': '028cefcbb71862a45dd1c0e11c1b17bbbb44987f6f9e813a3869e34f9044dec0'}

Transportation:

{'index': 2, 'previous_hash': '028cefcbb71862a45dd1c0e11c1b17bbbb44987f6f9e813a3869e34f9044dec0', 'current_transaction': {'transactions': [{'timestamp': 1660389278.193925, 'Name': 'Smartphone', 'Repairability': 'yes', 'Size': 'small', 'from': 'Transportation x', 'to': 'Collector x', 'Digital signature': 'approaved', 'msg': 'This waste is in good order', 'FLAG': 'N'}, {'timestamp': 1660389278.193925, 'Name': 'Tab', 'Repairability': 'yes', 'Size': 'small', 'from': 'Transportation x', 'to': 'Collector x', 'Digital signature': 'approaved', 'msg': 'This waste is in good order', 'FLAG': 'N'}, {'timestamp': 1660389278.193925, 'Name': 'MixerGrinder', 'Repairability': 'No', 'Size': 'medium', 'from': 'Transportation x', 'to': 'Collector x', 'Digital signature': 'approaved', 'msg': 'This waste is in good order', 'FLAG': 'N'}, {'timestamp': 1660389278.193925, 'Name': 'Laptop', 'Repairability': 'yes', 'Size': 'medium', 'from': 'Transportation x', 'to': 'Collector x', 'Digital signature': 'approaved', 'msg': 'This waste is in good order', 'FLAG': 'N'}, {'timestamp': 1660389278.193925, 'Name': 'TV', 'Repairability': 'No', 'Size': 'large', 'from': 'Transportation x', 'to': 'Collector x', 'Digital signature': 'approaved', 'msg': 'This waste is in good order', 'FLAG': 'N'}, {'timestamp': 1660389278.193925, 'Name': 'Desktop-Computer', 'Repairability': 'No', 'Size': 'large', 'from': 'Transportation x', 'to': 'Collector x', 'Digital signature': 'approaved', 'msg': 'This waste is in good order', 'FLAG': 'N'}]}, 'timestamp': 'datetime.now().timestamp()', 'nonce': 6, 'hash': '3546ef6056545b443f5c8ca5600383b80c76c480533b21ce5ea6a338bac1aff0'}

Collector:

{'index': 3, 'previous_hash': '3546ef6056545b443f5c8ca5600383b80c76c480533b21ce5ea6a338bac1aff0', 'current_transaction': {'transactions': [{'timestamp': 1660389278.193925, 'Name': 'Smartphone', 'Repairability': 'yes', 'Size': 'small', 'from': 'Collector x', 'to': 'Repairing shops x', 'Digital signature': 'approaved', 'msg': 'This waste is in good order', 'FLAG': 'N'}, {'timestamp': 1660389278.193925, 'Name': 'Tab', 'Repairability': 'yes', 'Size': 'small', 'from': 'Collector x', 'to': 'Repairing shops x', 'Digital signature': 'approaved', 'msg': 'This waste is in good order', 'FLAG': 'N'}, {'timestamp': 1660389278.193925, 'Name': 'MixerGrinder', 'Repairability': 'No', 'Size': 'medium', 'from': 'Collector x', 'to': 'Formal recycler x', 'Digital signature': 'approaved', 'msg': 'This waste is in good order', 'FLAG': 'N'}, {'timestamp': 1660389278.193925, 'Name': 'Laptop', 'Repairability': 'yes', 'Size': 'medium', 'from': 'Collector x', 'to': 'Repairing shops x', 'Digital signature': 'approaved', 'msg': 'This waste is in good order', 'FLAG': 'N'}, {'timestamp': 1660389278.193925, 'Name': 'TV', 'Repairability': 'No', 'Size': 'large', 'from': 'Collector x', 'to': 'Formal recycler x', 'Digital signature': 'approaved', 'msg': 'This waste is in good order', 'FLAG': 'N'}, {'timestamp': 1660389278.193925, 'Name': 'Desktop-Computer', 'Repairability': 'No', 'Size': 'large', 'from': 'Collector x', 'to': 'Formal recycler x', 'Digital signature': 'approaved', 'msg': 'This waste is in good order', 'FLAG': 'N'}]}, 'timestamp': 'datetime.now().timestamp()', 'nonce': 180, 'hash': '9de8f300f25fc52a7443420c270cdcc45e032fd96cdefb967b2d42788cc831c0'}

Formal recycler:

{'index': 4, 'previous_hash': '9de8f300f25fc52a7443420c270cdcc45e032fd96cdefb967b2d42788cc831c0', 'current_transaction': {'transactions': [{'timestamp': 1660389278.193925, 'Name': 'Metal', 'from': 'Formal recycler x', 'to': '3rd party metal recycler', 'Digital signature': 'approaved', 'msg': 'This waste is in good order', 'FLAG': 'N'}, {'timestamp': 1660389278.193925, 'Name': 'Plastic', 'from': 'Formal recycler x', 'to': '3rd party platic recycler', 'Digital signature': 'approaved', 'msg': 'This waste is in good order', 'FLAG': 'N'}, {'timestamp': 1660389278.193925, 'Name': 'Glass', 'from': 'Formal recycler x', 'to': '3rd party glass recycler', 'Digital signature': 'approaved', 'msg': 'This waste is in good order', 'FLAG': 'N'}]}, 'timestamp': 'datetime.now().timestamp()', 'nonce': 86, 'hash': '64493b84867695e48c0686660d297d4a1499aa877a985b8eb1d3614020372b50'}

Repairing_shops:

{'index': 5, 'previous_hash': '64493b84867695e48c0686660d297d4a1499aa877a985b8eb1d3614020372b50', 'current_transaction': {'transactions': [{'timestamp': 1660389278.193925, 'Name': 'Smartphone', 'Repairability': 'yes', 'Size': 'small', 'from': 'Repairing shops x', 'to': 'Refurbishing', 'Digital signature': 'approaved', 'msg': 'This waste is in good order', 'FLAG': 'N'}, {'timestamp': 1660389278.193925, 'Name': 'Tab', 'Repairability': 'yes', 'Size': 'small', 'from': 'Repairing shops x', 'to': 'Refurbishing', 'Digital signature': 'approaved', 'msg': 'This waste is in good order', 'FLAG': 'N'}, {'timestamp': 1660389278.193925, 'Name': 'Laptop', 'Repairability': 'yes', 'Size': 'medium', 'from': 'Repairing shops x', 'to': 'Refurbishing', 'Digital signature': 'approaved', 'msg': 'This waste is in good order', 'FLAG': 'N'}]}, 'timestamp': 'datetime.now().timestamp()', 'nonce': 133, 'hash': 'cc9b472b5b52421ad19bd4df527bfe844756d100fe0db5a4ce280e275a258b40'}

Refurbishing:

{'index': 6, 'previous_hash': 'cc9b472b5b52421ad19bd4df527bfe844756d100fe0db5a4ce280e275a258b40', 'current_transaction': {'transactions': [{'timestamp': 1660389278.193925, 'Name': 'Smartphone', 'Repairability': 'yes', 'Size': 'small', 'from': 'Refurbishing', 'to': '2nd hand market', 'Digital signature': 'approaved', 'msg': 'This waste is in good order', 'FLAG': 'N'}, {'timestamp': 1660389278.193925, 'Name': 'Tab', 'Repairability': 'yes', 'Size': 'small', 'from': 'Refurbishing', 'to': '2nd hand market', 'Digital signature': 'approaved', 'msg': 'This waste is in good order', 'FLAG': 'N'}, {'timestamp': 1660389278.193925, 'Name': 'Laptop', 'Repairability': 'yes', 'Size': 'medium', 'from': 'Refurbishing', 'to': '2nd hand market', 'Digital signature': 'approaved', 'msg': 'This waste is in good order', 'FLAG': 'N'}]}, 'timestamp': 'datetime.now().timestamp()', 'nonce': 14, 'hash': '7ead45e6130cb688bfb4440e090c9b63b447889a6ea6932540330e9c0a12c730'}

FIGURE 7.3 Output from the code.

digital technologies for supply chain digitization (Kosior 2022). Over the next five years, the SaaS-based SCM industry in Europe is anticipated to develop rapidly. The SCM market in Europe is expanding due to the end-user organizations' quick adoption of flexible and scalable digital technology for the efficient management of their supply chains (Businesswire.com 2022).

One of the major new technologies influencing Europe's future is blockchain. Blockchain can help to improve the efficiency of interactions between individuals, businesses, and government agencies, as well as strengthen trust and allow each party to keep custody of their own data. It will be critical in the development of a citizen-centric, sustainable, transparent, and inclusive European digital society. Investing in next-generation technologies such as blockchain will be critical to boosting Europe's technical sovereignty. According to the EU blockchain policy, 'The EU provides funding for blockchain research and innovation through grants and by supporting investment.' The EU and EEA Member States, along with the European Commission, have joined forces to create the European Blockchain Partnership (EBP), which aims to develop a Europe-wide blockchain infrastructure for international public services called the EBSI (European blockchain strategy, 2021). This should facilitate cross-border delivery of safer, distributed, and more citizen-centric public services among all subjects in EU member states (Hodžić & Owens, 2022). The EBP and EBSI will not only help familiarize EU officials with blockchain technology but also act as a sandbox for understanding technological use cases and adopting better policies (Valdavida et al. 2021).

Undoubtedly, the adoption of blockchain not only will be beneficial for developing a digital and transparent supply chain for EWM, but also will ensure the circularity of materials, conforming to the circular economy principles.

7.5 CONCLUSION

In this study, blockchain has been implemented in the EWM value chain with the aim to achieve a transparent and circular supply chain network. The work developed a blockchain based on the operational strategy of the supply chain. The chain is focused on identifying electronic items and components based on repairability and reusability at two different nodes, i.e., at the premises of collector-cum-dismantler and at the formal recycling unit. The developed programme works well with the given circumstances and ensures traceability of the supply chain until the third-party recyclers. Policy aspects from the European context were discussed in brief. Sustainability aspects of the digital e-waste supply chain were discussed with a focus on sustainable smart cities. However, there are many complex issues and challenges along the supply chain network of e-waste. Future studies involving issues of the informal sector and transboundary movement are underway with an aim to develop more sustainable versions of e-waste supply chain network.

ACKNOWLEDGEMENT

The authors acknowledge the support received from their respective institutes. Any opinions, findings, and conclusions expressed in this material are those of the author(s) and do not necessarily reflect the views of the affiliated universities or companies of the authors.

Contribution: Led by AD1, and co-conceptualized by BD, the foundation was substantiated by AD1 & AD2. Data mining and draft writing were carried out by all the authors. Coding was carried out by AD1 and AD2 with the help of AD3. SC was responsible for the European outlook. Sustainability analysis was carried out by BD inputs from AD1. Proofreading was carried out by AD3 and BD. All authors have read and agreed to this version of the manuscript.

REFERENCES

Ahmad, Raja Wasim, Khaled Salah, Raja Jayaraman, Ibrar Yaqoob, Mohammed Omar, and Samer Ellahham. "Blockchain-based forward supply chain and waste management for COVID-19 medical equipment and supplies." *IEEE Access* 9 (2021): 44905–44927.

Akram, Shaik Vaseem, Sultan S. Alshamrani, Rajesh Singh, Mamoon Rashid, Anita Gehlot, Ahmed Saeed AlGhamdi, and Deepak Prashar. "Blockchain enabled automatic reward system in solid waste management." *Security and Communication Networks* 2021 (2021).

Andreoni, Antonio, Justin Barnes, Anthony Black, and Timothy Sturgeon."Digitalization, industrialization, and skills development: opportunities and challenges for middle-income countries." (2021): 261–285.

Annamalai, Jayapradha. "Occupational health hazards related to informal recycling of E-waste in India: an overview." *Indian Journal of Occupational and Environmental Medicine* 19, no. 1 (2015): 61.

Baldé, C. P., Forti, V., Gray, V., Kuehr, R., & Stegmann, P. (2017). *The Global e-waste Monitor 2017: Quantities, Flows and Resources.* United Nations University, International Telecommunication Union, and International Solid Waste Association.

Brindhadevi, Kathirvel, Damià Barceló, Nguyen Thuy Lan Chi, and Eldon R. Rene. "E-waste management, treatment options and the impact of heavy metal extraction from e-waste on human health: Scenario in Vietnam and other countries." *Environmental Research* (2022): 114926.

Businesswire.com. "Europe SAAS SCM Market Report 2022: Driven by acceleration of supply chain digitization, e-commerce boost and EU regulations on data storage – researchandmarkets.com." Business Wire, August 2, 2022. Available at: https://www.businesswire.com/news/home/20220802005663/en/Europe-SaaS-SCM-Market-Report-2022-Driven-by-Acceleration-of-Supply-Chain-Digitization-E-Commerce-Boost-and-EU-Regulations-on-Data-Storage---ResearchAndMarkets.com. (Accessed: October 25, 2022).

Chaudhary, Karishma, Prasanna Padmanabhan, Deepak Verma, and Prashant Dev Yadav. "Blockchain: a game changer in electronic waste management in India." *International Journal of Integrated Supply Management* 14, no. 2 (2021): 167–182.

Debnath, Biswajit. "Sustainability of WEEE recycling in India." In *Re-Use and Recycling of Materials*, pp. 15–32. River Publishers, 2022.

Debnath, Biswajit. "Towards sustainable e-waste management through industrial symbiosis: a supply chain perspective." In *Industrial Symbiosis for the Circular Economy*, pp. 87–102. Springer, Cham, 2020.

Dias, Pablo, Md Tasbirul Islam, Bin Lu, Nazmul Huda, and Andréa M. Bernarde. "e-waste transboundary movement regulations in various jurisdictions." *Electronic Waste: Recycling and Reprocessing for a Sustainable Future* (2022): 33–59.

European blockchain strategy – brochure. (no date). "Shaping Europe's digital future." Available at: https://digital-strategy.ec.europa.eu/en/library/european-blockchain-strategy-brochure (Accessed: October 25, 2022).

Hodžić, Sabina, and Jeffrey Owens. "Policy note: blockchain technology: potential for digital tax administration." *Intertax* 50, no. 11 (2022).

Kaza, Silpa, Lisa Yao, Perinaz Bhada-Tata, and Frank Van Woerden. *What a Waste 2.0: A Global Snapshot of Solid Waste Management to 2050*. World Bank Publications, 2018.

Khan, Atta Ur Rehman, and Raja Wasim Ahmad. "A blockchain-based IoT-enabled e-waste tracking and tracing system for smart cities." *IEEE Access* 10 (2022): 86256–86269.

Kosior, Katarzyna. "The advancement of digitalization processes in food industry enterprises in the European Union." (2022).

Orlins, Sabrina, and Dabo Guan. "China's toxic informal e-waste recycling: local approaches to a global environmental problem." *Journal of Cleaner Production* 114 (2016): 71–80.

Parajuly, Keshav, Khim B. Thapa, Ciprian Cimpan, and Henrik Wenzel. "Electronic waste and informal recycling in Kathmandu, Nepal: challenges and opportunities." *Journal of Material Cycles and Waste Management* 20, no. 1 (2018): 656–666.

Rejeb, Abderahman, Karim Rejeb, Steve Simske, and Horst Treiblmaier. "Blockchain technologies in logistics and supply chain management: a bibliometric review." *Logistics* 5, no. 4 (2021): 72.

Rosário, Albérico Travassos, and Joana Carmo Dias. "Sustainability and the digital transition: a literature review." *Sustainability* 14, no. 7 (2022): 4072.

Saberi, Sara, Mahtab Kouhizadeh, Joseph Sarkis, and Lejia Shen. "Blockchain technology and its relationships to sustainable supply chain management." *International Journal of Production Research* 57, no. 7 (2019): 2117–2135.

Sahoo, Swagatika, and Raju Halder. "Blockchain-based forward and reverse supply chains for e-waste management." In *International Conference on Future Data and Security Engineering*, pp. 201–220. Springer, Cham, 2020.

Sahoo, Swagatika, Arnab Mukherjee, and Raju Halder. "A unified blockchain-based platform for global e-waste management." *International Journal of Web Information Systems* (2021).

Spieske, Alexander, Maximilian Gebhardt, Matthias Kopyto, and Hendrik Birkel. "Improving resilience of the healthcare supply chain in a pandemic: Evidence from Europe during the COVID-19 crisis." *Journal of Purchasing and Supply Management* (2022): 100748.

Terazono, Atsushi, Masahiro Oguchi, Aya Yoshida, Ruji P. Medina, and Florencio C. Ballesteros. "Material recovery and environmental impact by informal e-waste recycling site in the Philippines." In *Sustainability through Innovation in Product Life Cycle Design*, pp. 197–213. Springer, Singapore, 2017.

Valdavida, Magdalena Cordero, Mirko Iaconisi, and Spyridon Pilos. "Is public sector ready for blockchain? Is blockchain ready for the public sector?." *European Review of Digital Administration & Law* 2, no. 2 (2021): 7–17.

Wath, Sushant B., Atul N. Vaidya, P. S. Dutt, and Tapan Chakrabarti. "A roadmap for development of sustainable E-waste management system in India." *Science of the Total Environment* 409, no. 1 (2010): 19–32.

Widmer, Rolf, Heidi Oswald-Krapf, Deepali Sinha-Khetriwal, Max Schnellmann, and Heinz Böni. "Global perspectives on e-waste." *Environmental Impact Assessment Review* 25, no. 5 (2005): 436–458.

Part 3

Industrial Perspective and Resilience for Future

8 Upcycling of E-waste

A Smart Approach Towards Sustainability

Anil Potluri Chowdary and Sayana Sai Brahmani

8.1 INTRODUCTION

Upcycling generally refers to the conversion of waste materials to something useful or valuable. This concept can be applied not only to the waste design industry but also to waste recycling and resource flow (Herman et al. 2018). Our study highlights upcycling as the key concept for improving the value of waste by redefining the concept as 'the recycling of waste materials and discarded products in ways that strengthen their value.'

Higher standards of living and technological development, especially in the electric and electronic industry, have shortened the product cycle of electrical and electronic products, leading to an overall increase of waste electronic and electrical equipment (WEEE), widely known as e-waste (Debnath et al. 2016). At the same time, the rise in the prices of raw metal has brought considerable attention to the recovery of valuable metals, such as copper, zinc, nickel, etc., included in e-waste (Debnath et al. 2022). Many countries like Denmark, Germany, and Korea have recently introduced a recycling target management system for e-waste and set a target for recycling per capita (6.0 kg per person) to be met by 2018. These developments have increased the demand for research on resource flow and recycling, especially concerning the promotion of e-waste recycling (Debnath et al. 2023).

One concept that has been gaining importance in e-waste recycling is upcycling. Upcycling is a combination of the words 'upgrade' and 'recycling' and is presently used to refer to the creation of products with added value using recycled materials. Our study focuses on upcycling as the key concept for improving the value of e-waste by refining the broad use of the concept into a practically applicable definition and suggesting strategies to promote upcycling based on domestic and foreign case studies. Then, the material flows of e-waste are analyzed, specifically on valuable resources and low-value residues that can be obtained from discarded refrigerators and computers, including the quantities of recovered resources and final throughputs. Based on the results of the material flow analysis (MFA),

DOI: 10.1201/9781003317050-11

we examine the available upcycling technologies for recovering and recalculating resources from e-waste and review how upcycling can be more effectively promoted in terms of material flow.

8.2 MATERIALS AND METHODOLOGY

Here we look specifically at e-waste upcycling. Most everyday prospective waste can be recycled, but they are other ways to reuse it. These include items required for motherboards of CPUs, dead earphones, and essential materials for making items.

8.2.1 Collection

Collection of waste and used products is not something new. Consumers habitually throw away products after usage. Traditionally, collection of these products was done by scavengers and rag pickers. Recently, formal models have been built by firms to recover waste and used products. This can be attributed to the increasing awareness of their harmful effects on the environment (Paras and Curteza 2018). In developed countries, it is mainly done by charity organizations, whereas in developing countries it is still done by local vendors who exchange the e-waste for groceries or buy them.

8.2.2 Material Sorting

E-waste is collected from different sources like industries, offices, households, etc. Initially, the material is tested to see whether it can be used again or should be dismantled. Usually from households you could see different types of electronic items where most of them look good. Instead of dismantling them, we can upcycle them. So, here comes the challenging part of reusing them. Instead of throwing them in plastic, which has minimal value, upcycling them into a new product adds an impressive price to it. Likewise, technology should be developed to make upcycling more efficient and cost effective.

8.2.3 Upcycling

Recycling may be considered the use of the material properties. However, several scholars proposed various definitions of upcycling. These include the following:

- Through upcycling value/quality of the product is improved by making a superior product (Dervojeda et al., 2014; Singh 2022).
- Giving new value to materials that are either discarded or are not being used anymore (Fletcher and Grose, 2012).
- Repurposing lower-value items such as a neck scarf to create a higher-value end use item, such as a wrap skirt or halter top (Janigo and Wu, 2015; Paras et al. 2019).

8.2.4 Redesigning

Redesigning may be considered one of the important steps in the upcycling process. The process of redesign is to add value to discarded or used products. The original idea of the redesign is based on the technique of pattern making and draping. The extent of redesign can vary from adding minor details to complete transformation of any item we choose (Paras et al. 2016; Paras 2019). Materials and design always differ from the type of material and type of upcycling object being made.

8.3 RESULTS AND DISCUSSION

Upcycling is one of the best ways of reusing e-waste parts into a new object or item. Here, four cases of e-waste upcycling are presented.

8.3.1 Case 1: Jewellery from E-waste

In Figure 8.1, motherboards are cut into required sizes and pierced with ear hangings to make jewellery. Earphones are also upcycled to make designer earrings. Capacitors are removed from different motherboards and are designed to make bracelets. Instead of throwing them away, they can be made into beautiful jewellery.

8.3.2 Case 2: CD and Cassette Upcycling

Figure 8.2 shows CDs cut into different sizes and attached on cardboard to make a photo frame. CDs are designed to make fish structures as well. In the bottom row of Figure 8.2, different cassettes are collected and painted; from those cassettes, tape is removed and used for tape art. The cassette's tape is used to make wall hangings and cartoon pieces that are painted.

8.3.3 Case 3: Desktop Upcycling

Figure 8.3 shows a desktop mouse and water bottle and soda caps designed as a butterfly wall hanging or showcase piece. A printer scanner, when dismantled,

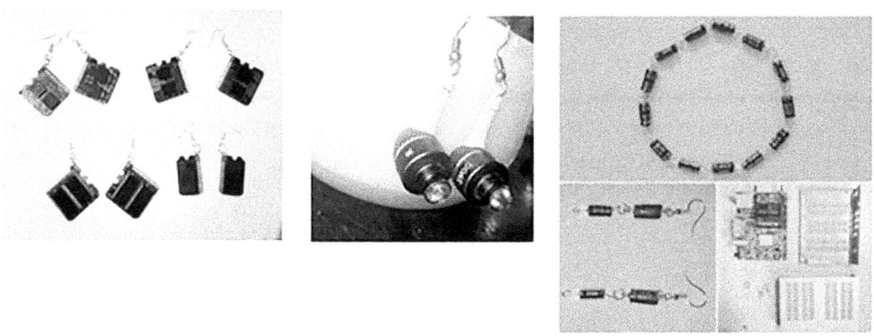

FIGURE 8.1 Upcycled jewellery from e-waste.

FIGURE 8.2 Upcycled products from CDs and cassettes.

provides fibre glass and plastic, but if you upcycled it by painting it, it turns into beautiful wall art or a portrait. Upcycling by painting is one of the best ways to reuse plastic. Floppy discs are the oldest parts of the CPU; these are excellent items for transforming into pen stands. They can also be used as succulent plant holders. The last photo, a tree, is designed using a CPU motherboard. The branches are designed by using iron strings that are removed from fibre net wires, and the motherboard itself has holes in it. These wires are tangled and created as branches, and decorative lights are arranged on them to give a glowing look. It can be used as a night light anywhere in the house.

8.3.4 Case 4: Upcycling of RAM and Bulbs

Figure 8.4 shows a chopper designed out of RAMs and IT waste. RAMs are designed as wings, and plastic is used as blades and rollers. This can be used as a decorative item. The second photo shows CLF bulbs; when you dismantle them,

FIGURE 8.3 Upcycled motherboard, mouse, scanner, etc.

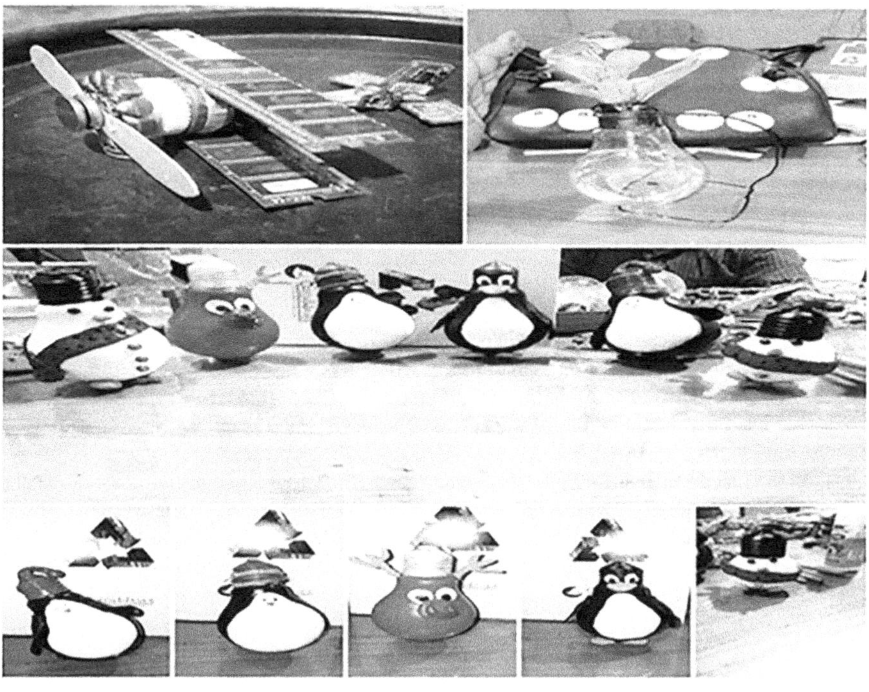

FIGURE 8.4 Upcycled RAM and bulbs.

you end up getting glass and filament, so CLF bulbs are designed as penguins, snowmen, etc. Using an iron string, the bulb is hung as a plant holder.

8.4 CONCLUSION

Upcycling is not a new thing. It is quite an old practise with a new name. In countries like India, it is often termed 'work education,' which is part of the school curriculum. Hence, the practise needs reviving for both economic development and sustainability. Yes, handling e-waste can be tricky and needs proper safety training. The cases presented in this chapter show the veracity of upcycling that can be applied to develop products for multiple purposes. Indeed, upcycling e-waste is a sustainable process; however, there might be some debates regarding the numbers. More case studies in this area will be helpful for upcycling e-waste and stopping it from going to landfills. Additionally, LCA studies should consider upcycling as an important route for better understanding of sustainability.

REFERENCES

Debnath, Biswajit, Roychoudhuri, Reshma, & Ghosh, Sadhan K. (2016). E-Waste Management – A Potential Route to Green Computing. Procedia Environmental Sciences, 35, 669–675 10.1016/j.proenv.2016.07.063.

Debnath, Biswajit, Das, Ankita, & Das, Abhijit (2022). Towards circular economy in e-waste management in India: Issues, challenges, and solutions, Circular Economy and Sustainability (pp. 523–543). 10.1016/b978-0-12-821664-4.00003-0.

Debnath, Biswajit, Das, Abhijit, Chowdary, Potluri Anil, & Bhattacharyya, Siddhartha (2023). Development in E-waste Management. 10.1201/9781003301899

Dervojeda, K., D. Verzijl, E. Rouwmaat, L. Probst, and L. Frideres. "Clean technologies: circular supply chains", Business Innovation Observatory, European Union, (2014), September 2014, available at: http://ec.europa.eu/DocsRoom/documents/13396/attachments/3/translations

Fletcher, K., and L. Grose. *Fashion & Sustainability: Design for Change*, Laurence King Publishing, London, 2012.

Herman, Krzysztof, Madalina Sbarcea, and Thomas Panagopoulos. "Creating green space sustainability through low-budget and upcycling strategies." *Sustainability* 10, no. 6 (2018): 1857.

Janigo, K. A., and J. Wu. (2015), "Collaborative redesign of used clothes as a sustainable fashion solution and potential business opportunity." *Fashion Practice* 7, no. 1, pp. 75–98.

Paras, Manoj Kumar, A. Curteza, and Rudrajeet Pal. "A State-of-the-Art Literature Review of Upcycling: A Clothing Industry Perspective." In *16th Romanian Textiles and Leather Conference, Romania, October 27–29, 2016*, p. 121. Editura Acreditata de Cncsis Bucuresti, 2016.

Paras, Manoj Kumar, and Antonela Curteza. "Revisiting upcycling phenomena: a concept in clothing industry." *Research Journal of Textile and Apparel* 22, no. 1 (2018): 46–58.

Paras, Manoj Kumar, Antonela Curteza, and Geetika Varshneya. "Identification of best reverse value chain alternatives: a study of Romanian used clothing industry." *Journal of Fashion Marketing and Management: An International Journal* (2019).

Singh, Jagdeep. "The sustainability potential of upcycling." *Sustainability* 14, no. 10 (2022): 5989.

9 Advanced Green Approaches for E-waste Recycling with Resource Recovery

Preetiman Kaur and Shivani Sharma

9.1 INTRODUCTION

The electronics industry has flourished as the biggest industry on the globe today. Thousands of new electronic devices, ranging from mobile phones and washing machines to WiFi modems and air conditioners, are manufactured every day. With the development of more innovative technologies, electronics production is diversifying and multiplying with a rapid pace, thus producing a large amount of e-waste (Dave et al., 2016). This huge increment in e-waste generation is due to the disposal of electronic equipment, either in pursuit of updated machinery or the end of a product's life.

Developed countries export e-waste to countries such as India and Bangladesh, where they waste recycle, which ultimately results in the accumulation of enormous e-waste (Agnihotri, 2011). This has turned out to be a health hazard for the labourers involved in dismantling and recycling. Dismantling involves unscrewing, shredding, tearing, and burning. Circuits are burnt to extract the precious and valuable metal ions such as gold, platinum, cadmium, and copper. The composition of e-waste includes various hazardous heavy metals, non-degradable plastics, acids, and toxic chemicals (Awasthi et al., 2016; Awasthi, 2018). Recycling of valuable elements (gold, silver, copper) found in e-waste has now become an effective way of acquiring income specifically in the informal sector of many developing countries. On the other side, recycling techniques expose labourers to a large number of hazardous and toxic substances (Manomaivibool, 2009). Moreover, the burning of the wire coat of these electronic devices consists of polyvinylchloride (PVC) and polychlorinated biphenyls (PCB), which produce hazardous smoke and carbon particles on burning. This release of hazardous substances may lead to lung and skin cancer (Sadiku et al., 2016). As a result, e-waste has become a global concern. A United Nation (UN) report points out that global e-waste is expected to exceed 40 million tonnes per year. In most of the countries, the recycling of electronic end products leads to the release of heavy pollution; it is non-cost effective and unregulated. Globally, the amount of

DOI: 10.1201/9781003317050-12

e-waste produced is expected to reach 52.2 million tonnes in 2021, in comparison to 44.7 million tonnes in 2016, with a yearly rise of approximately 3–4% (Balde et al., 2017). Not all e-waste finishes the recycling process. A large amount of it is dumped in land-fills, which leads to consequential health disorders due to leaching of toxic metals into the water table, which ultimately makes its approach towards people through agricultural products.

To save the overall ecosystem from getting contaminated, there is a dire need to look for alternatives for heavy metals/toxic metals remediation present in e-waste. Different physico-chemical approaches such as incineration, landfill dumping, chemical oxidation, reduction, adsorption, and precipitation have been adopted to minimize the effects of e-waste, but these methods have their own drawbacks and limitations (Al-Garni, 2005). To overcome these limitations and drawbacks, there are biological approaches that are considered green chemistry approaches for the extraction of toxic metals from e-waste, which are modest, environment-loving, and reliable methods (Iqbal et al., 2019). The biological approach includes the transformation of toxic components to non-toxic forms by microorganisms that involve the elimination of natural and man-made pollutants through natural bio-processes (Sivasubramanian, 2006). These microorganisms own efficient metabolic machinery that employs metals as an energy source for both growth as well as development processes, electron acceptors during respiration (aerobic), fermentation (anaerobic), and co-metabolism (Igiri et al., 2018). Hence, in this chapter different green biological approaches will be discussed for the proper handling of e-waste and the recovery of metal ions.

9.2 COMPOSITION OF E-WASTE

Precious/hazardous and/or toxic components that are present initially in e-waste are referred to as primary contaminants. These include gold, platinum, palladium, cadmium, lead, tin, antimony, mercury, nickel, cadmium, chromium, copper, arsenic, and barium. Other metals present in parts per million (ppm) ranges are selenium, americium, gallium, beryllium, etc. These metal species are being used in smoke detectors, semiconductors, and rectifiers and have some side effects. Metal ions temper metabolism and are considered a carcinogen because they can cause cancerous effects when inhaled as dust or smoke. Heavy dose exposure to selenium is also dangerous as it can cause selenosis. In addition, rare earth metal species are being used in the production of printed circuit boards (PCBs), cathode ray tubes (CRTs), and to provide toughness and thermal qualities of different alloys in equipment batteries (Liu et al., 2009). There are different types of PCBs on the basis of the material used. FR1, FF2, FR3, FR4 (FR = flame retardant), and so on are the major examples. FR1 (130°C) has higher glass transition (TG) temperature than FR2 (105°C). FR3 contains epoxy resin binder instead of phenolic resin. FR4 (TG = 120°C –130°C) is mostly used in the industry; it's a glass epoxy fibre laminated material.

Exposure to rare earth metals increases the risk of many diseases, specifically respiratory bronchitis. This category also includes organic pollutants like polychlorinated biphenyls, polybrominated diphenyl ethers (PBDEs), tetrabromobisphenol-A

TABLE 9.1
Comparison of Different Metallurgy Processes

Parameters	Pyrometallurgy	Hydrometallurgy	Biohydrometallurgy
Impact on environment	Very high due to emissions of harmful gases	Moderate, as toxic chemicals are being released	Much less
Selectivity	Much smaller amount of metals are being selected	More in comparison to pyrometallurgy	More in comparison to other approaches
Economics	Very high capital is required	Less capital is required	Very low investment is required
Process conditions	Very harsh conditions are there	Corrosive compounds are being produced	Very few noxious compounds are being released

(TBBPA), polybrominatedBiphenyls (PBBs), hexabromocyclododecane (HBCD), and flame retardants. The exposure to these contaminants may also lead to different types of cancers in humans.

The second type of contaminant in e-waste is secondary contaminants. Secondary contaminants are the products obtained once the e-waste has been processed during the recovery and extraction of valuable components. Different physico-chemical methods (Table 9.1), such as mechanical (Meng et al., 2017), pyrometallurgical (Jung et al., 2017), hydrometallurgical (Pant et al., 2012), and biometallurgical (Mrazikova et al., 2016) processes, are generally used for extracting the metal ions from PCBs. Physico-chemical methods that are used for the extraction of heavy metals usually release toxic harmful gases such as dioxins, which drastically affect nature as well as public health.

9.3 GREEN APPROACHES: BIOLEACHING

Bioleaching is known to be an effective biological method. In this process, microorganisms are used to extract metal ions from PCBs (e-waste) with a low-cost and environmentally friendly method (Hong and Valix, 2014). The main mechanism of using bacterial cultures involves the generation of the major ore oxidizer (e.g. oxidation of ferric iron). This process occurs on the bacterial cell membrane. During this process, bacterial cells attain energy as the generation of free electrons, which are employed for reducing the oxygen to produce water exothermically, which hence liberates energy and is used for further metabolic activities. It includes the sulphide oxidation by generated ferric (Fe^{2+}) ion along with the regeneration of ferric ion reactants.

Most commonly chemolithoautotrophs (*Acidithiobacillus ferrooxidans, Leptospirillum ferrooxidans*) are employed to extricate different elemental species from PCBs (Erüst et al., 2013). Many fungal cultures such as *Aspergillus niger* and *Penicillium simplicissimum* also have the ability to dissolute heavy metals from the contaminated site, but the leaching mechanism is somewhat different

(Erüst et al., 2013). Fungal cultures secrete their acids, which they often produce during their metabolic activities to dissolve the metal from the polluted site.

There are two major types of mechanisms of bioleaching: direct and indirect bioleaching. Generally, the indirect bioleaching process is most accepted by researchers all over the globe (Liu et al., 2017). The indirect bioleaching process is further divided into contact and non-contact biooxidation/bioleaching mechanisms (Jadhav and Hocheng, 2013).

Contact mechanism: In this process, bacteria get attached to the surface and produce a biofilm, which leads to the accumulation of biomass, which provides the interface of reaction at the site of contamination. Biofilm is the actual site of oxidation of Fe^{2+} by bacteria to Fe^{3+}, and further, the surface is dissolved by Fe^{3+} (Liu et al., 2017). It has been demonstrated that the production of biofilm by bacterial secretions plays a significant role in bioleaching (More et al., 2014). This is because the amount of reduced iron (Fe) and hydrogen ion in biofilm is much greater in comparison to that in solution, which helps in increasing the rate of dissolution of acids at contaminated sites (ores) (Coram and Rawlings, 2002).

Non-contact mechanism: In this way of bioleaching, the bacteria are not adhered to the ore/contaminated surface. The role of the oxidized form of iron (Fe^{3+}) produced by bacteria is to induce the dissolution of the targeted surface. Out of these mechanisms, the indirect contact bioleaching mechanism is generally known as the major bioleaching process of PCBs (Silva et al., 2015), as represented in Figure 9.1.

Further, to supplement the mechanism of the bioleaching process, biooxidation of copper metal ions from PCBs is depicted in the following paragraph:

This process is bifurcated (copper (Cu) ions dissolution via acid production from PCBs) into two phases. The first step involves oxidation of ferrous ions by bacterial culture to form ferric ions, whereas the second step of this process involves the most prominent step of this mechanism in which copper is mobilized from the PCBs by the ferric ions from the previous step and the ferric ions are again reduced to ferrous ions. Hence, there is occurrence of a cyclic redox reaction between oxidized and reduced forms of metal ions, which ultimately

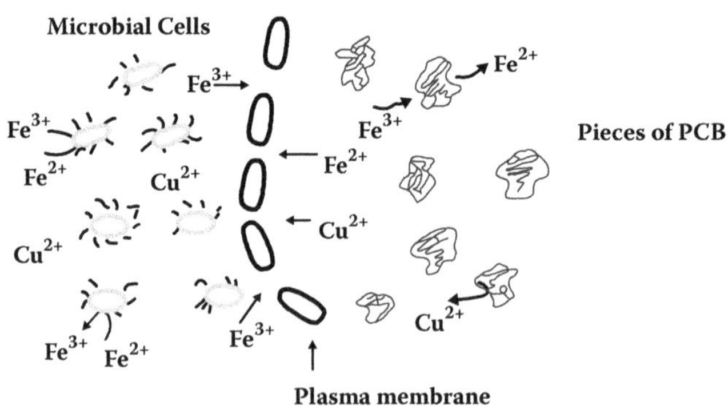

FIGURE 9.1 *Acidithiobacillus ferrooxidans* bacterial adhesion pictorial representation on crushed PCB during the process of bioleaching (adapted from Silva et al., 2015).

results in leaching of copper from the site (Yang et al., 2009). Many researchers have investigated the mechanism of bioleaching of copper from PCBs (e-waste; Arshadi and Mousavi, 2015; Mrážiková et al., 2016). The toxic components in PCBs disturb the growth and metabolic activities of bacteria, which is concluded to be one of the main reasons for the low amount of leaching of metal ions (Yang et al., 2014). One of the researchers (Bryan et al., 2015) demonstrated that the metabolic activity of bacteria might be suppressed by the presence of different metal species (Cu^{2+}, Sn^{2+}, Cr^{3+}, Ni^{1+}, Zn^{2+}, and Cu^{2+}). PCBs contain some non-metallic components that sometimes increase the toxicity related to bacteria in the biooxidation process (Zhu et al., 2011). The reduction of the toxicity of PCBs includes the following equation:

$$Fe(1)^{2+} + O_2 + 4H^+ \text{ Bacteria}^- \rightarrow Fe^{3+} + 2H_2O$$
$$2Fe(2)^{3+} + Cu \rightarrow 2Fe^{2+} + Cu^{2+}$$

In one of the studies, bacteria are documented for treating the accumulated e-waste (Faramarzi et al., 2004). Actinobacteria belongs to the category of both metabolically and ecologically active bacterial cultures that are present in huge numbers specifically in metal-contaminated sites. Hence, in one of the studies, *Acidithiobacillus ferrooxidans* was being used to oxidize ferrous iron (Fe^{2+}) and to generate reduced ferric iron (Fe^{3+}), which causes the solubilization of metal species. Further, the microorganisms also result in the production of reduced metal ions such as oxidized iron and/or H^+ from reduced iron/sulphur, which is used for transforming the insoluble Cu, Zn, and Ni metals to form their respective water-soluble Cu^{2+}, Zn^{2+}, and Ni^{2+} sulphates (Liang et al., 2010). Cyanogenic bacteria have the capability to catalyze the formation of cyanide ion (CN) along with hydrocyanic acid (HCN) as their secondary metabolites, which plays a significant role in solubilising the precious earth metals like gold (Au), silver (Ag), and platinum (Pt) (Natarajan and Ting, 2014). This is how bioleaching, an eco-friendly process, plays an important role in providing a cost-effective and easy way to extract the valuable metals from e-waste.

9.4 BIOSORPTION

Biosorption is known as quick and reversible form of ion-binding method as it is not dependent on cellular metabolism, such that a dead biomass can be employed (Volesky, 1994). There are many advantages of biosorption; in this process, there is a chance to restore or reuse the adsorbents (microbial cultures) for a long run, with no noxious effects of elevated concentrations of metals along with no limitations of nutrients (Gadd, 2009). Through this biological process, metals can be captured readily and desorbed. Moreover, this biological method can be performed over a wide range of temperatures (4–90°C) and pH (3–9) (Abbas et al., 2014). Different functional groups present in microbial biomass usually help in different processes of adsorption; only because of this cause, metal species as well as their concentrations can be retained via using a different variety of biosorbents (Volesky, 2003).

9.4.1 Microbes as Biosorbents

9.4.1.1 Bacteria

Morphology of bacteria consists of the cell wall, the outer-most layer; the cell membrane, the penultimate layer; the capsule; and the slime layer. These are the major coverings of the bacterial cell, which contains different types of functional groups that are involved in adorption, such as the carboxyl group (COO^-), the amino group (NH_2), phosphate (PO_4), and some sulphate groups (SO_4). Presence of functional groups helps in chelation of metal species from the polluted site (mentioned in Figure 9.2). There are different mechanisms involved in metal binding at the cell wall surface of microbial cultures (Kanamarlapudi et al., 2018). Bacteria possess an adequate biosorption capability because of their high surface-to-volume ratio as well as ubiquitous nature; hence, they can easily grow in extreme environmental conditions.

9.4.1.2 Algae

The outer-most covering of autotrophic microorganism algae mainly constitutes alginic acid, the chitin layer, certain xylan, and mannans with effective functional groups such as imidazole, sulphate, and hydroxyl moiety, which provides a surface for the attachment of certain metal ions (Abbas et al., 2014).

9.4.1.3 Fungi

The cell wall of fungi constitutes approximately 90% of polysaccharides with varied functional groups in their outer covering such as uranic acids, ligands containing nitrogen moiety, carboxyl, and phosphate groups (Huang and Huang, 1996). These moieties help in adsorbing the metal species present in e-waste. Biosorption/passive uptake of different metal ions occurs as a metabolism-independent process; hence, fungal biomass can act as an efficient biosorbent further from which recovery of metal species will be effective and eco-friendly (Shamim, 2018). This energy-independent pathway of uptake of metal species is generally influenced

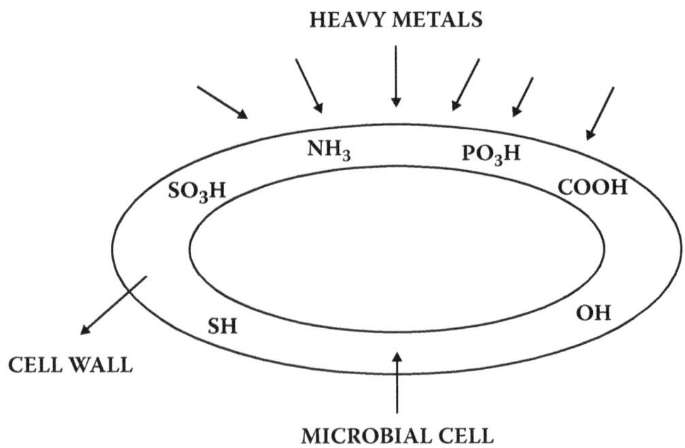

FIGURE 9.2 Pictorial representation of chelation of metal ions via cell wall functional groups.

by temperature and certain metabolic inhibitors, etc. Mushrooms, the macrofungi, are found in different types of habitats such as forests, contaminated soils, or water bodies. Their different morphological parts help in proper uptake of metal ions. A network of hyphae known as mycelium, certain germinating structures such as sporocarps, participate effectively in capturing ions from the contaminated sites (Abbas et al., 2014).

9.4.1.4 Yeast

Another microbial category includes the yeast known as *Saccharomyces cerevisiae*, which is regarded as the best model system for the analysis of biosorption due to its certain advantages as it can easily grow, is avirulent in nature, and provides high biomass yield while growing in a simple growth medium (Kapoor and Viraraghavan, 1997). The efficiency of biosorption via yeast cells depends on several factors in terms of the valency and radius of metal ions, the age of the yeast culture, and the composition of the medium in which the yeast has been grown (Table 9.2). Additionally, the substantial size of yeast helps in making it a favourable microbe for the bioremediation of metal ions (Mapolelo et al., 2004).

9.4.2 Use of Biosorbent Immobilization

The traditional way of using free as well as dead microbial biomass as one of the best biosorbents in a continuous system usually has a number of demerits, including

TABLE 9.2
Different Metals Absorbed by Different Microorganisms

Bacteria	Metal Ions Biosorbed	References
Pseudomonas spp.	Cr, Hg, Cd, and Pb	Alia et al., 2016
Bacillus thuringiensis strain OSM29	Ni, Cu, Cd	Oves et al., 2013
Ralstonia pickettii	Mn	Huang et al., 2018
Kocuria spp.*BRI 36*	Ni and Cr	Mulik et al., 2018
Fungus		
Paecilomyces marquandii	Zn and Pb	Xia et al., 2015
Termitomyces clypeatus	Cr	Ramrakhiani and Majumder, 2011
Mucor rouxii and Aspergillus niger	Pb, Cd, Ni, and Zn	Yan and Viraranhavan, 2008
Beauveria bassiana and Metarhizium anisopliae	Cd and Pb	Hussein et al., 2011
Verticillium insectorum J3	Pd and Zn	Feng et al., 2018
Colletotrichum spp.	Cd and Pb	Silva et al., 2015
Algae		
Lyngbya taylorii	Pb, Ni, Cd, and Zn	Wilke et al.,2006
Anabaena sphaerica	Cd and Pb	Abdel et al., 2013
Ulva lactuca	Cu, Cr, Cd, and Pb	Ibrahim et al., 2016

difficulty in separating the biomass, wastage of biosorbent after regeneration, and low strength during adherence along with low rigidity. Immobilization through biopolymeric matrix is one such technique to overcome the disadvantages of free microbial biomass. Such immobilization techniques help in improving the biosorbent performance by maintaining proper mechanical strength and defending against certain chemical attacks (Hao and Liu, 2009). Two major forms of immobilization (entrapment and encapsulation) can be applied for immobilizing microbes (Volesky, 2001). These matrices are mostly composed of polyacrylamide, polysulphone, alginate, and polyurethane types of polymers (Bai and Abraham, 2003).

One of the researchers revealed that the use of live as well heat-inactivated fungal culture, *Trametes versicolor,* was capable of immobilizing within the beads of carboxyl methyl cellulose (CMC) and was capable of removing copper, lead, and zinc from the medium efficiently. Both live and heat-inactivated biosorbents showed appreciable biosorption capacity (1.51 and 1.84 mmol for copper, 0.85 and 1.11 mmol for lead, and 1.33 and 1.67 mmol for zinc, respectively). Hence these results showed that the model is the best fit for Langmuir isotherm (Bayramoğlu et al., 2003).

One of the significant aspects of biosorption is that it is a surface process that involves the interaction between metal and biosorbent by electrostatic interactions and Vander Waal forces. Therefore, this process seems to be easy for the recovery of metals due to weak bonding (Chojnacka, 2010). The biosorption process generally depends on physicochemical properties of metals, their concentration, their molecular weight, their ionic radius, and their oxidation state.

Many bacterial and fungal species are being used for the biosorption of certain metal ion species, such as cadmium, chromium, lead, and uranium, from e-waste.

9.4.3 Bioaccumulation

In an ecosystem transformation, decomposition and bioaccumulation of various heavy metals are mainly due to different activities performed by microorganisms, although when these metal species are present in higher concentrations, they react to produce noxious compounds in cells (Nies, 1999). Microbes also possess a capacity to convert different toxic ions into insoluble compounds that one can easily dispose off (Rodriguez et al., 1993). There are certain mechanisms of microbes in order to overcome metal-stressed conditions (Schmidt et al., 2013). The majority comprise the efflux (excretion) of elemental ions outside the cell as well as the accumulation and complex formation of the metal ions in the interior of the cell, along with reducing (redox reaction) of the metal species into a less toxic state. In comparison to other physico-chemical methods, biological approaches are considered more environmentally friendly and cost-effective. When living cells accumulate metals in active mode, it is known as bioaccumulation that depends on different physiological, intrinsic, and genetic adaptations, structural and biochemical properties, and metal specification, along with the ability to obtain a noxious effect of metal ions. Thus, it is elucidated as the utilization of toxicants by metabolism-dependent cells followed by their transport into the cell for further metabolic activities. As it is dependent on the growth of an organism, it is arbitrated only to living biomass (Jaafar et al., 2015), in contrast to biosorption, which can occur through dead biomass.

In a study by Aderonke et al. (2017), the most promising microorganisms that were tolerant to zinc, copper, and lead were isolated and screened from soils that were contaminated with metal ions due to the disposal of e-waste, and further, these microbial cultures were analyzed for their metal ions by their capability for accumulation. The results of this study showed that samples from different soil isolates that were screened from an e-waste dumpsite had more microbes than those from the other two sites that were not dumped with e-waste. This study reflected that 17 bacteria and five fungi showed very high tolerance to Zn, Cu, and Pb. Although all bacterial and fungal isolates have the capacity to accumulate different heavy metals with varying rates ranging from 200 to 3,200 mg kg^{-1}.

9.4.4 Biomineralization

E-waste recycling is a very significant aspect not only for managing the waste, but also for recovering and extracting certain valuable metals. Biomineralization is one of the microbial processes in which toxic metal ions present in a contaminated site bind with ions or ligands that are produced from the microorganisms either on the cell wall surface or secreted from the cell during certain metabolic activities to perform precipitation (Achal et al., 2012).

In one of the studies by Das and Ting (2018), two-step bioleaching of e-waste was conducted using an engineered bacterium *Chromobacterium violaceum* followed by biomineralization of gold nanoparticles. From this study, gold was recovered from the bioleachate via a biomineralization process with the help of the bacterium *Delftia acidovorans,* which performed the process by converting reduced soluble gold to form nanoparticles. Approximately 68.7% of gold was recovered by using this biological process.

9.4.5 Biotransformation

Biotransformation is defined as a mechanism in which a compound is transformed from one type of chemical structure to another more stable compound via undergoing certain chemical reactions; while transforming toxic metal species to non-toxic species, the oxidized state transformed by adding the electrons, which leads to change in their chemical properties (Prakash et al., 2012). Two major ways are present for the occurrence of this process.

1. *Direct reduction (enzymes involved):* Multivalent noxious elemental species undergo reduction. This process occurs in the exterior of the cell from the enzyme activity.
2. *Indirect reduction:* Multivalent toxic metal ions are reduced and immobilized via the action of the bacterial culture, which has the capacity to reduce the ions in the contaminated site (Tabak et al., 2005).

9.5 RECOVERY OF METALS

9.5.1 After Biosorption

To achieve the cost-effective extraction of precious and valuable metal species after metal uptake, it is necessary to regenerate the biosorbent (Kiran et al., 2005). This regeneration of metal ions is known as desorption; its objective includes owning the adsorption capability of the specific biosorbent (Alluri et al., 2007). Desorption process is to be performed in such a way that an efficient amount of metal should be obtained in the concentrated form as well as the original state of the biosorbent should not suffer in order to reuse the capability of the biosorbent (Ahemad and Kibret, 2013). Therefore, there is a need to use an eco-friendly chemical that is harmless to the microbial biomass and also provides the effective metal-binding capacity. Here becomes the role of green chemistry, which has an objective to employ a sustainable and environmentally friendly approach to reduce pollution and to recover as well recycle useful material out of the waste.

Eluents (dilute mineral acids) such as sulphuric acid, nitric acid, hydrochloric acid, and certain organic acids such as lactic acid and nitric acid, along with different complexing agents like ethylenediaminetetraacetic acid (EDTA) and thiosulphate, are being used for the biosorbent and for metal recovery. There was complete desorption of chromium (VI) metal from the microbial biomass of *Mucor Hiemalis* by using eluent NaOH (0.1 N). The biomass was capable of retaining biosorption and desorption capacity for up to almost five cycles. As it was also investigated that this result fit well with the Langmuir isotherm statistic model and its Fourier Transform Infrared, spectroscopy FTIR analysis demonstrated the presence of amino groups, which helped in biosorption (Figure 9.3) (Tewari et al., 2005).

9.5.2 After Bioleaching

The failure of the bioleaching process to uptake different elements (silver, lead) from matrices is due to the presence of silver (Ag) as a re-calcitrant sulphide.

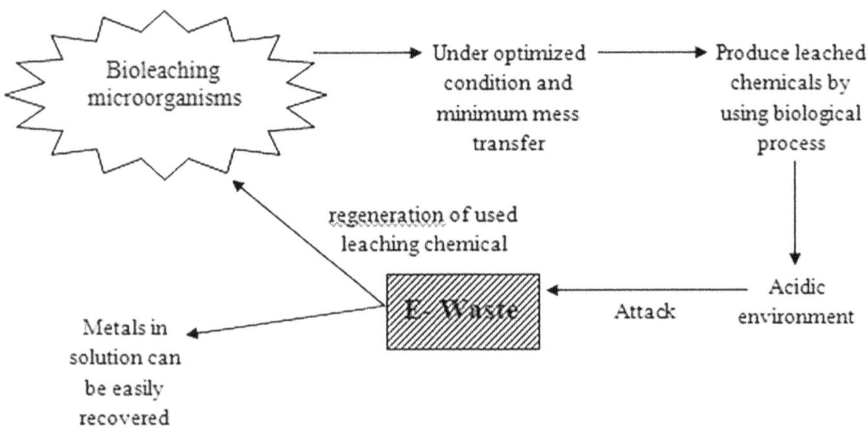

FIGURE 9.3 Recovery of metals from e-waste via bioleaching.

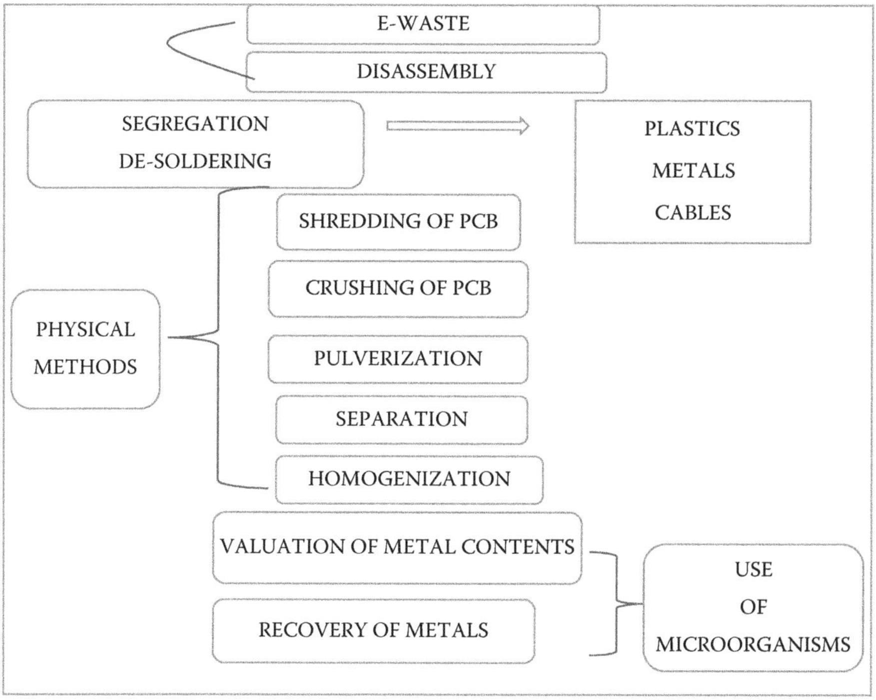

FIGURE 9.4 Flow chart representing the different methods involving in disassembly of e-waste.

In the case of Pb, lead sulphate is formed by oxidation of lead, whose solubility is much less for recovering of metal species present in the residues.

Therefore, in one of the studies by Sabrina Hedricha et al. (2018), after the process of bioleaching, hot brine leaching was used, by keeping the fact that silver and lead elements are able to complex with chloride (Cl^-) in acidic solutions, e.g.:

$$PbSO_4 + 4NaCl \rightarrow Na(PbCl)NaSO_4$$
$$AgCl + NaCl \rightarrow Na(AgCl)$$

Different concentrations of NaCl can be used with varied solution pH (adjusted with HCl); the process of leaching was performed remarkably better when the assays were performed at pH 0 than for those whose pH was adjusted to 2 or without any pH adjustment (Figure 9.4).

9.5.3 Green Chemistry: Management and Control of E-waste

Regardless of the development potential of information and communication technologies (ICTs), the discarding of different noxious e-wastes causes a serious tenability challenge (Passerini and Wu, 2008) as e-waste becomes toxic if not discarded properly. Waste management and control hierarchy involve the following methods to achieve a sustainable approach (Agunwamba, 2001): source reduction and waste recycling, which involves reclaiming and reuse of maximum waste. Green chemistry/sustainable chemistry works towards reducing and preventing pollution. Sustainable chemistry involves research in the chemical industry and engineering, which encourages the product's design and methods, which can deflate the employment as well as production of harmful substances (Wikipedia, 2010, 'Green chemistry,' Retrieved on 6/5/2010 from http://en.wikipedia.org/wiki/Green_chemistry).

The usage of chemicals leads to reduced production of waste products and noxious compounds and ultimately improves the efficiency of the process (US EPA (United States Environmental Protection Agency) (2010). It is a very advantageous way to prevent pollution as it involves contemporary scientific ideas to overcome environmental problems. Chemicals used in this approach are generally safe for both humans as well as for environmental applications. Hence, these chemicals do not tamper with the ecosystem equilibrium, and they are non bio-accumulative in nature and inherently easy to handle (US EPA (United States Environmental Protection Agency) (2010).

9.6 CONCLUSION

E-waste management is a huge and complex task due to improper implementation of disposal practises. This chapter discusses the global and Indian outline of e-waste production along with its noxious effects on the human health and environment equilibrium. Different biological approaches, such as biosorption, bioleaching, and bioaccumulation, can be used over the physico-chemical methods to minimize adverse environmental consequences. Biological methods utilize organisms' machinery and biomass to extract metal ions rather than producing lethal gas fumes and acids, which further contaminate the environment robustly. The smaller size of microbes provides a large area-to-volume ratio, and their cell wall composition is found to be an effective tool for adsorbing different metal ions in an eco-friendly manner.

Physico-chemical and biochemical ways of recovering a metal from e-waste have their own pros and cons. There are certain effective, technical, economical, scientific, and environmental reasons for selecting one process in place of other. A hybrid of these techniques has the capability to lift the stumbling blocks associated with a particular approach. This idea is helpful in exploring an effective and new area of biometallurgy that can help in the recovery of precious metal species that are present in trace quantities in their respective ores via using metabolic activities of microorganisms. Various valuable metal recycling from e-waste as well as removal of secondary resources from e-waste can be successful through these innovative techniques. Hence, this can be proved effective in the recycling of e-waste through sound management.

REFERENCES

Abbas, S. H., I. M. Ismail, T. M. Mostafa and A. H. Sulaymon. 2014. Biosorption of heavy metals: A review. *Journal of Chemical Science and Technology* 3(4):74–102.

Abdel-Aty, A.M., N.S. Ammar, H.H. Abdel Ghafar and R.K. Ali. 2013. Biosorption of cadmium and lead from aqueous solution by fresh water alga Anabaena sphaerica biomass. *Journal of Advanced Research* 4:367–374.

Achal, V., Pan, X., Fu, Q. and Zhang, D. 2012. Biomineralization based remediation of As(III) contaminated soil by Sporosarcina ginsengisoli. *Journal of Hazardous Materials*, 201-202:178–184.

Aderonke, A.K., Oladimeji, O.O., Shittu, O.B., Okeyode, I.C. and Taiwo, M.O. 2017. Bioaccumulation of heavy metals using selected organisms isolated from electronic waste dumpsite of two south-western states in Nigeria. *Applied Environmental Research*, 39(2):29–40.

Agnihotri, V. 2011. In: *E-waste in India. Research Unit in Larrdis, Rajya Sabha Secretariat*, Rajya Sabha, Delhi, India. 2011; pp. 1–127.

Agunwamba, J.C. 2001. *Waste Engineering and Management Tools*, Enugu: Immaculate Publications.

Ahemad, M. and M. Kibret. 2013. Recent trends in microbial biosorption of heavy metals: A review. *Biochemistry and Molecular Biology* 1(1):19–26.

Al-Garni, S. M. 2005. Biosorption of lead by Gram -ve capsulated and non-capsulated bacteria. *Water SA* 31(3): 345–349.

Alia, A., M. Sihabudeen and Z. Hussain. 2016. Biosorption of heavy metals by Pseudomonas bacteria. *International Research Journal* of *Engineering* and *Technology* 3: 1446–1450.

Alluri, H. K., S. R. Ronda, V. S. Settalluri, J. S. Bondili, V. Suryanarayana and P. Venkateshwar. 2007. Biosorption: An eco-friendly alternative for heavy metal removal. *African Journal of Biotechnology* 6:2924–2931.

Arshadi, M. and S. M. Mousavi. 2015. Multi-objective optimization of heavy metals bioleaching from discarded mobile phone PCBs: Simultaneous Cu and Ni recovery using Acidithiobacillus ferrooxidans. *Separation and Purification Technology* 147:210–219.

Awasthi, A. K., X. Zeng and J. Li. 2016. Environmental pollution of electronic waste recycling in India: A critical review. *Environmental Pollution* 211: 259–270.

Awasthi, A. K., F. Cucchiella, I. D'Adamo, J. Li, P. Rosa, S. Terzi, G. Wei and X. Zeng. 2018. Modelling the correlations of e-waste quantity with economic increase. *Science of the Total Environment* 613-614, 46–53.

Bai, R. S. and T. E. Abraham. 2003. Studies on chromium (VI) adsorption-desorption using immobilized fungal biomass. *Bioresource Technology* 87:17–26.

Baldé, C. P., V. Forti, V. Gray, R. Kuehr and P. Stegmann. 2017. *The Global E-waste Monitor*, United Nations University (UNU), International Telecommunication Union (ITU) & International Solid Waste Association (ISWA), Bonn/Geneva/Vienna.

Bayramoğlu, G., S. Bektaş and M. Y. Arıca. 2003. Biosorption of heavy metal ions on immobilized white-rot fungus Trametes versicolor. *Journal of Hazardous Materials* 101:285–300.

Bryan, C. G., E. L. Watkin, T. J. McCredden, Z. R. Wong, S. T. L. Harrison and A. H. Kaksonen. 2015. The use of pyrite as a source of lixiviant in the bioleaching of electronic waste. *Hydrometallurgy* 152:33–43.

Chojnacka, K. 2010. Biosorption and bioaccumulation-the prospects for practical applications. *Environment International*. 36:299–307.

Coram, N. J. and D. E. Rawlings. 2002. Molecular relationship between two groups of the genus Leptospirillum and the finding that Leptospirillum ferriphilum sp. nov. Dominates South African commercial biooxidation tanks that operate at 40°C. *Applied and Environmental Microbiology* 68:838–845.

Das, S. and Y.-P. Ting. 2018. Sonobioleaching of E-waste and biomineralization of gold nanoparticles (online). In: *Chemeca 2018*. Queenstown: Institution of Chemical Engineers, 2018: 101.1-101.10. Availability: <https://search.informit.com.au/documentSummary;dn=048451430087906;res=IELENG> ISBN: 9781911446682. (cited 16 Aug 20).

Dave, S. R., M. B. Shah and D. R. Tipre. 2016. E-Waste: Metal pollution threat or metal resource? *Journal of Advanced Research in Biotechnology* 1(2): 14. Page 2 of 14 10.15226/2475-4714/1/2/00103

Erüst, C., A. Akcil, C. S. Gahan, A. Tuncuk and H. Deveci. 2013. Biohydrometallurgy of secondary metal resources: A potential alternative approach for metal recovery. *Journal of Chemical Technology & Biotechnology* 88: 2115–2132.

Faramarzi, M. A., M. Stagars, E. Pensini, W. Krebs and H. Brandl. 2004. Metal solubilization from metal-containing solid materials by cyanogenic Chromobacterium violaceum. *Journal of Biotechnology* 113:321–326.

Feng, Cl., J. Li, X. Li, Kl. Li and K. Luo. 2018. Characterization and mechanism of lead and zinc biosorption by growing Verticillium insectorum J3. *PLoS One* 13(12): 10.1371/journal.pone.0203859

Gadd, G. M. 2009. Biosorption: Critical review of scientific rationale, environmental importance and significance for pollution treatment. *Journal of Chemical Technology and Biotechnology* 84:13–28.

Hao, Z. L. and Y. M. Liu. 2009. Bioleaching of heavy metals from sewage sludge. *Journal of Shijiazhuang Vocational Technology Institute*. 2:005.

Hedricha, S., R., Kermerb, T., Aubelb, M., Martinb, A., Schippersa, D., Barrie Johnsonc and E., Janneckb. 2018. Implementation of biological and chemical techniques to recover metals from copper-rich leach solutions. *Hydrometallurgy* 179, August 2018: 274–281.

Hong, Y. and M. Valix. 2014. Bioleaching of electronic waste using acidophilic sulfur oxidising bacteria. *Journal of Cleaner Production* 65: 465–472.

Huang, C. and C. P. Huang. 1996. Application of Aspergillus oryzae and Rhizopus oryzae for Cu (II) removal. *Water Research* 30:1985–1990.

Huang, H., Y. Zhao, Z. Xu, Y. Ding and W. Zhang. 2018. Biosorption characteristics of a highly Mn(II)-resistant Ralstonia pickettii strain isolated from Mn ore. *PLoS One* 13(8): 10.1371/journal.pone.0203285

Hussein, K. A., S.H. Hassan and J. H. Joo. 2011. Potential capacity of Beauveria bassiana and Metarhizium anisopliae in the biosorption of Cd2+ and Pb2. *Journal of General and Applied Microbiology* 57(6), 347–355. 10.2323/jgam.57.347

Ibrahim, W. M., F. H. Asad and A. A. Yahia 2016. Biosorption of toxic heavy metals from aqueous solution by Ulva lactuca activated carbon. *Egyptian Journal of Basic and Applied Sciences* 3,241–249.

Igiri, B. E., S. I. R. Okoduwa, G. O. Idoko, E. P. Akabuogu, A. O. Adeyi and I. K. Ejiogu. 2018. Toxicity and bioremediation of heavy metals contaminated ecosystem from Tannery wastewater: A review. *Journal of Toxicology* 2018, Article ID 2568038:16.

Iqbal, A, M. Manzoor and M. Arshad. 2019. Biodegradation of e-waste pollution. *Electronic Waste Pollution*, 291–306. 10.1007/978-3-030-26615-8_19.

Jaafar, R., A. BioAl-Sulami, A. Al-Taee, F. Aldoghachi and S. Napes. 2015. Biosorption and bioaccumulation of some heavy metals by Deinococcus radiodurans isolated from soil in Basra Governorate-Iraq. *Journal of Biotechnology and Biomaterial* 5(2): 1–5.

Jadhav, U. and H. Hocheng. 2013. Extraction of silver from spent silver oxide-zinc button cells by using Acidithiobacillus ferrooxidans culture supernatant. *Journal of Cleaner Production* 44:39–44.

Jung, M., K. Yoo and R. D. Alorro. 2017. Dismantling of electric and electronic components from waste printed circuit boards by hydrochloric acid leaching with stannic ions. *Materials Transactions* 58: 1076–1080.

Kapoor, A. and T. Viraraghavan. 1997. Fungi as biosorbent. In: DAJ W, Forster CF, editors. *Biosorbents for Metal Ions*. London: Taylor & Francis; pp. 67–85.

Kiran, I., T. Akar and S. Tunali. 2005. Biosorption of Pb (II) and Cu (II) from aqueous solution by pretreated biomass of Neurospora crassa. *Process Biochemistry* 40(11): 3550–3558. 10.1016/j.procbio.2005.03.051

Liang, G., Y. Mo and Q. Zhou. 2010. Novel strategies of bioleaching metals from printed circuit boards (PCBs) in mixed cultivation of two acidophiles. *Enzyme and Microbial Technology* 47:322–326.

Liu, J., W. Wu, X. Zhang, M. Zhu and W. Tan. 2017. Adhesion properties of and factors influencing Leptospirillum ferriphilum in the biooxidation of refractory gold-bearing pyrite. *International Journal of Mineral Processing* 160: 39–46.

Liu, Q., K. Q. Li, H. Zhao, G. Li and F. Y. Fan. 2009. The global challenge of electronic waste management. *Environmental Science and Pollution Research* 16(3): 248–249.

Manomaivibool, P. 2009. Extended producer responsibility in a non-OECD context: the management of waste electrical and electronic equipment in India. *Resources, Conservation and Recycling* 53(3): 136–144. 10.1016/j.resconrec.2008.10.003.

Mapolelo, M. and N.Torto. 2004. Trace enrichment of metal ions in aquatic environments by Saccharomyces cerevisiae. *Talanta* 64:39–47. 10.1016/j.talanta.2003.10.058

Meng, L., Z. Wang, Y. W. Zhong, L. Guo, J. T. Gao, K. Y. Chen, H. J. Cheng and Z. C. Guo. 2017. Supergravity separation for recovering metals from waste printed circuit boards. *Chemical Engineering Journal* 326: 540–550.

More, T. T., J. S. Yadav, S. Yan, R. D. Tyagi and R. Y. Surampalli. 2014. Extracellular polymeric substances of bacteria and their potential environmental applications. *Journal of Environmental Management* 144:1.

Mrazikova, A., J. Kadukova, R. Marcincakova, O. Velgosova, J. Willner, A. Fornalczyk and M. Saternus. 2016. The effect of specific conditions on Cu, Ni, Zn and Al recovery from PCBS waste using acidophilic bacterial strains. *Archives of Metallurgy and Materials* 61:261–264.

Mrážiková, A., J. Kaduková, R. Marcinčáková, O. Velgosová, J. Willner, A. Fornalczyk and M. Saternus. 2016. The effect of specfifc conditions on Cu, Ni, Zn and Al recovery from PCBS waste using acidophilic bacterial strains. *Archives of Metallurgy and Materials* 61:261–264.

Mulik, A. R., P. Kulkarni and R. K. Bhadekar. 2018. Biosorption Studies on Nickel and Chromium by Kocuria spp. BRI 36 Biomass. *International Journal of Applied Engineering Research* 13, 6886–6893.

Natarajan, G. and Y. P. Ting. 2014. Pretreatment of e-waste and mutation of alkali-tolerant cyanogenic bacteria promote gold biorecovery. *Bioresource Technology* 152: 80–85.

Nies, D. H. 1999. Microbial heavy-metal resistance. *Applied Microbiology and Biotechnology* 51: 730–750.

Oves, M., M.S. Khan and A. Zaidi. 2013. Biosorption of heavy metals by Bacillus thuringiensis strain OSM29 originating from industrial effluent contaminated north Indian soil. *Saudi Journal of Biological Sciences* 20,121–129.

Pant, D., D. Joshi, M. K. Upreti and R. K. Kotnala. 2012. Chemical and biological extraction of metals present in E waste: a hybrid technology. *Waste Management* 32: 979–990.

Passerini, K. and D. Wu. 2008. The new dimension of collaboration: Mega and intelligent communities. *ICT and Wellbeing Journal of Knowledge Management* 12(5): 79–90.

Prakash, B., M. Debnath and G. B. K. S. Prasad. 2012. Microbes: Concepts and applications. 10.1002/9781118311912.index.

Ramrakhiani, L. and K. S. Majumder. 2011. Removal of hexavalent chromium by heat inactivated fungal biomass of Termitomyces clypeatus: Surface characterization and mechanism of biosorption. *Chemical Engineering Journal* 171(3), 1060–1068.

Ramya Krishna Kanamarlapudi, S. L., Kumar Chintalpudi, V. and Muddada, S. 2012. Application of biosorption for removal of heavy metals from wastewater. 10.5772/intechopen.77315

Rodriguez, M. L., R. L. C. de la Cruz, R. N. Farias and E. M. Massa. 1993. Membrane-associated redox cycling of copper mediates hydroperoxide toxicity in Escheri-chia coli. *Bio-chemistry Biophysical Acta* 1144: 77–84.

Sadiku, M. N. O., S. M. Musa and S. R. Nelatury. 2016. What E-waste is all about. *Journal of Scientific and Engineering Research* 3(5):128–130.

Schmidt, I., O. Sliekers, M.C. Schmid, E. Bock, J. Fuerst, J.G. Kuenen and M. Strous. 2013. New concepts of microbial treatment processes for the nitrogen removal in wastewater. *FEMS Microbiological Reviews* 27: 481–492.

Shamim, S. 2018. Biosorption of Heavy Metals. Biosorption of Heavy Metals 10.5772/intechopen.72099

Silva, R. A., J. Park, E. Lee, J. Park, S. Q. Choi and H. Kim. 2015. Influence of bacterial adhesion on copper extraction from printed circuit boards. *Separation and Purification Technology* 143:169–176.

Sivasubramanian, V. 2006. Phycoremediation - issues and challenges. *Indian Hydrobiol* 9(1): 13–22.

Tabak, H., P. N. L. Lens, E. van Hullebusch and W. Dejonghe. 2005. Developments in Bioremediation of soils and sediments polluted with metals and radionuclides – 1. microbial processes and mechanisms affecting bioremediation of metal contamination and influencing metal toxicity and transport. *Reviews in Environmental Science and Bio/Technology* 4. 115–156. 10.1007/s11157-005-2169-4.

Tewari, N., P. Vasudevan and B. Guha. 2005. Study on biosorption of Cr (VI) by Mucor hiemalis. *Biochemical Engineering Journal* 23:185–192.

US EPA (United States Environmental Protection Agency). 2010. Introduction to the concept of Green Chemistry. Retrieved on 6/5/2010 from http://www.epa.gov/greenchemistry/pubs/about_gc.html

Volesky, B. 1994. Advances in biosorption of metals: Selection of biomass types. *FEMS Microbiology Reviews* 14:291–302.

Volesky, B. 2001. Detoxification of metal-bearing effluents: Biosorption for the next century. *Hydrometallurgy* 59:203–216.

Volesky, B. 2003. Biosorption process simulation tools. *Hydrometallurgy* 71:179–190.

Wikipedia, the free encyclopedia. 2010. "Green chemistry," Retrieved on 6/5/2010 from http://en.wikipedia.org/wiki/Green_chemistryLtd

Wilke, A., R. Buchholz and G. Bunke. 2006. Selve biosorption of heavy metals by algae. *Environmental Biotechnology* 2, 47–56.

Xia, L., X. Xu, W. Zhu, Q. Huang and W. Chen. 2015. A comparative study on the biosorption of Cd2+ onto Paecilomyces lilacinus XLA and Mucoromycote sp. *International Journal of Molecular Sciences* 16, 15670–15687.

Yan, G. and T. Viraraghavan. 2008. Mechanism of biosorption of heavy metals by Mucor rouxii. *Engineering in Life Sciences* 8, 363–371.

Yang, T., Z. Xu, J. Wen and L. Yang. 2009. Factors influencing bioleaching copper from waste printed circuit boards by Acidithiobacillus ferrooxidans. *Hydrometallurgy* 97: 29–32.

Yang, Y., S. Chen, S. Li, M. Chen, H. Chen and B. Liu. 2014. Bioleaching waste printed circuit boards by Acidithiobacillus ferrooxidans and its kinetics aspect. *Journal of Biotechnology* 173:24–30.

Zhu, N., Y. Xiang, T. Zhang, P. Wu, Z. Dang, P. Li and J. Wu. 2011. Bioleaching of metal concentrates of waste printed circuit boards by mixed culture of acidophilic bacteria. *Journal of Hazardous Materials* 192:614–619.

10 Conclusion

Biswajit Debnath, Abhijit Das, Anil Potluri, and Siddhartha Bhattacharyya

Electronic waste (e-waste) has created chaos worldwide. Due to its hazardous nature, it is often shipped to other developing and underdeveloped nations. In the recent past, the Philippines declared war over the issue of e-waste against Canada (The Guardian 2019). In many developed countries the burden of e-waste is ignored. China banned the import of waste materials, including plastic waste and e-waste, from other developed countries. As a result, waste disposal has become a problem for these countries. The global e-waste generation in 2016 is equivalent to nearly 4,600 Eiffel towers (Balde et al. 2017), which is a pretty overwhelming number. Generation of e-waste grew exponentially and reached 53.6 million metric tonnes in 2019; it is expected to reach 120 million tonnes by 2050 (Forti et al. 2020). India is the third largest e-waste producer in the world, generating nearly 2 million tons per year. To tackle this huge amount of e-waste, immediate attention is required in terms of technological advancement and management aspects (Debnath et al. 2022).

E-waste is a huge source of secondary raw materials. It is a heterogeneous material, and it contains an assortment of materials, including metals, polymers, and siliceous materials, including glass (Debnath 2022). This makes e-waste a potential candidate to enhance urban mining and help in establishing a circular economy (Debnath et al. 2022). The term 'urban mining' is synonymous with resource recovery from e-waste. Such discarded e-waste streams thus trap a huge amount of metallic and non-metallic resources, which relegates urban mining of e-waste as not an option anymore, but rather a necessity (Cossu, 2015; Debnath et al. 2022). Material recovered from e-waste could be a feedstock for several other allied industries, which can also bring up industrial symbiotic models (Debnath 2020). E-waste is a complex material, and the issues coming up from it are also very complex. While the hardware part of e-waste poses a health hazard to the informal e-waste workers, improper disposal of memory devices can pose security threats at different levels. With the advancement in reverse engineering technologies, such threats are becoming more prominent (Roychowdhury et al. 2019). Another emerging issue that is less explored is uncertainties along the e-waste supply chain. This is an important aspect that needs immediate attention for sustainable e-waste management (Pedro et al. 2023). To combat all these issues pertaining to e-waste management, it is important to strategize in advance.

DOI: 10.1201/9781003317050-13

In the past two decades, there has been tremendous research and development in the e-waste management sector. Technological advancements have generated several routes for urban mining of e-waste, including mechanical recycling, hydro-metallurgical, pyro-metallurgical technologies, etc. (Alam et al. 2022). These can be designated as e-waste recycling and valorization technologies. On the other hand, the mushrooming and disruptive information technologies, such as cloud computing, edge computing, Internet of Things (IoT), and blockchain, are leading the paradigm shift towards a more intelligent and sustainable e-waste management system. These can be termed 'e-waste supervision technologies.' While it is important to dive deeper into the development of the technologies for urban mining of e-waste, it is also relevant to give equal importance to these new information technologies that hold the potential to make the supply chain not only digital, but also smart and traceable.

A sustainable e-waste management system is essential for a digital economy spanning smart villages and sustainable smart cities. This cannot be achieved with exclusive implementation of these two ranges of technologies. To achieve that, an optimal intervention of e-waste valorization and recycling technologies, policy tools, and e-waste supervision technologies, such as digital technologies, is essential. Additionally, evolving green recycling technologies need to be backed up for transformation from 'Lab-to-land.' Hence, the paradigm shift can only happen with the adaptation of the new technologies and their implementation to build intelligent, smart, and sustainable systems. This volume may come up as a handbook in identifying the recent advancements and technological proliferations leading towards building an intelligent and sustainable society, and it may help researchers, e-waste entrepreneurs, and supply chain managers understand the well-times paradigm shift in the e-waste management sector.

REFERENCES

Alam, Tanvir, Rabeeh Golmohammadzadeh, Fariborz Faraji, and M. Shahabuddin. "E-Waste Recycling Technologies: An Overview, Challenges and Future Perspectives." *Paradigm Shift in E-waste Management* (2022): 143–176.

Baldé, Cornelis P., Vanessa Forti, Vanessa Gray, Ruediger Kuehr, and Paul Stegmann. *The Global E-waste Monitor 2017: Quantities, Flows and Resources*. United Nations University, International Telecommunication Union, and International Solid Waste Association, 2017.

Cossu, Raffaello. "Urban Mining: Concepts, Terminology, Challenges." *Waste Management* 45 (2015): 1–3.

Debnath, Biswajit, Ankita Das, and Abhijit Das. "Towards Circular Economy in E-waste Management in India: Issues, Challenges, and Solutions." In *Circular Economy and Sustainability*, pp. 523–543. Elsevier, 2022.

Debnath, Biswajit. "Sustainability of WEEE Recycling in India." In *Re-Use and Recycling of Materials*, pp. 15–32. River Publishers, 2022.

Debnath, Biswajit. "Towards Sustainable E-waste Management through Industrial Symbiosis: A Supply Chain Perspective." In *Industrial Symbiosis for the Circular Economy*, pp. 87–102. Springer, Cham, 2020.

Forti, Vanessa, Cornelis P. Balde, Ruediger Kuehr, and Garam Bel. *The Global E-waste Monitor 2020: Quantities, Flows and the Circular Economy Potential*. United Nations

University (UNU)/United Nations Institute for Training and Research (UNITAR) – co-hosted SCYCLE Programme, International Telecommunication Union (ITU) & International Solid Waste Association (ISWA), Bonn/Geneva/Rotterdam, 2020.

Roychowdhury, P., J. M. Alghazo, Biswajit Debnath, S. Chatterjee, and O. K. M. Ouda. "Security Threat Analysis and Prevention Techniques in Electronic Waste." In *Waste Management and Resource Efficiency*, pp. 853–866. Springer, Singapore, 2019.

Senna, Pedro, Lino G. Marujo, Augusto da Cunha Reis, and Ana de Souza Gomes dos Santos. "Circular E-Waste Supply Chains' Critical Challenges: An Introduction and a Literature Review." *Conversion of Electronic Waste into Sustainable Products* (2023): 233–250.

The Guardian. "Trash Talk: Philippine President to 'Declare War' on Canada in Waste Dispute." *The Guardian. Guardian News and Media*, 2019. April 24. https://www.theguardian.com/world/2019/apr/24/philippine-president-rodrigo-duterte-to-declare-war-on-canada-in-waste-dispute.

Index

For Product Safety Concerns and Information please contact our EU
representative GPSR@taylorandfrancis.com
Taylor & Francis Verlag GmbH, Kaufingerstraße 24, 80331 München, Germany

www.ingramcontent.com/pod-product-compliance
Ingram Content Group UK Ltd.
Pitfield, Milton Keynes, MK11 3LW, UK
UKHW021508240926
475702UK00035B/656

* 9 7 8 1 0 3 2 3 2 8 5 6 0 *